Brian Minter's
— New —
Gardening
— Guide —

Brian Minter's New Gardening Guide

Whitecap Books
Vancouver / Toronto

Copyright © 1998 by Brian Minter and Greg Rasmussen
Whitecap Books
Vancouver/Toronto

All rights reserved. No part of this publication may be reproduced, stored in a retrieval system, or transmitted in any form or by any means, electronic, mechanical, photocopying, recording or otherwise, without prior written permission of the publisher.

The information in this book is true and complete to the best of our knowledge. All recommendations are made without guarantee on the part of the author or Whitecap Books Ltd. The author and publisher disclaim any liability in connection with the use of this information. For additional information please contact Whitecap Books Ltd., 351 Lynn Avenue, North Vancouver, BC V7J 2C4.

Edited by Elaine Jones
Technical edit by Carolyn Jones
Proofread by Elizabeth McLean
Illustrations by Warren Clark
Cover photo by Paddy Wales (Audrey Litherland's Garden); inset photo by Ryan McNair
Cover and interior design by Tanya Lloyd

Printed and bound in Canada

Canadian Cataloguing in Publication Data

Minter, Brian, 1947–
    Brian Minter's new gardening guide

    Includes index.
    ISBN 1-55110-624-8

    1. Gardening—Canada. I. Rasmussen, Greg. II. Title.
III. Title: New gardening guide.
SB451.36.C3M56 1998   635.9'0971   C98-910012-X

The publisher acknowledges the support of the Canada Council and the Cultural Services Branch of the Government of British Columbia in making this publication possible. We acknowledge the financial support of the Government of Canada through the Book Publishing Industry Development Program for our publishing activities.

**For more information on this and other Whitecap Books titles,
please visit our web site at www.whitecap.ca.**

To my wife, Faye, who came back to work with me late at night so many times after long, hard days to offer encouragement and support. Her constant feedback and suggestions made this a better book. And to our two daughters, Lisa and Erin, who make each day a delight.

Acknowledgements

In a book of this nature, you often have second thoughts about the choices of plants you recommend to others. You also question your objectivity when you have a passion for plants.

There are so many great folks in the gardening world upon whom I rely for feedback and information. I must, however, learn to phone them before midnight—I've found their sense of humour and interest in plants is much less when you wake them from a dead sleep! The many old-time gardeners I meet daily and who always have a gardening story to share have been a source of learning, as have the many wonderful listeners to my radio program who continually push me to improve my knowledge. I deeply appreciate the flow of kind letters and e-mails that provide information when I honestly don't know the answers.

When it comes to evaluating plants objectively, I rely on John Derrick to temper my enthusiasm. Fred Wein has been a constant source of information on clematis. The pruning advice in his booklet *Clematis,* I follow to the T. Angus Richardson, Don Biggin and Dave Wall have always kept me abreast of what's new in lawns. David Wilson is my guide on mini roses and heather. Marian Vaughan has taught me a great deal about old roses, rhodies and perennials, and Geunther Boch has always been a terrific help in evaluating new perennials. Paul and Nick Reimer have kept me up to date on Japanese maples and magnolias, and Dr. Todd has been my magnolia connoisseur. Ken Wilson has been helpful with botanical information and critiquing as has Elke Knechtel.

Greg Rasmussen deserved a gold medal for translating my illegible handwriting and tapes of my speed-talking radio show into readable stuff. Greg and I have done this book together and I am so grateful for all his patience and great style. Jo-Ann Chabot deserves the same for not only translating my handwriting, but getting it out fast.

There are many others who deserve mention for their kind support and encouragement.

Contents

Introduction 9

Getting to the Root of It 11

Making the Most of Your Vegetable Patch 25

Herbs and Edible Flowers 49

Bedding Plants 59

Hanging Baskets and Planters 71

Perennials 85

Roses 103

Vines and Climbers 117

Deciduous Trees and Shrubs 127

Broadleaf Evergreens and Conifers 147

Fruit and Nut Trees and Small Fruits 163

Bulbs 193

Lawns and Other Groundcovers 203

Index 216

Introduction

In responding to gardening questions on open-line radio shows over the past 20 years, I have noticed an increased frustration about where to turn for advice on everything from fruit trees and lawns to roses, perennials, herbs, bulbs, rhododendrons and much more. Your questions have prompted this book.

Many gardeners, especially those who are new to this wonderfully creative art, simply want advice about the most reliable performers in every plant family. If they are going to take time out of a busy schedule and expend the energy to plant something, they want results. The goal of this book is to suggest some of the top-performing plants in many different areas of gardening.

I love all plants, and I'm a bit of a fanatic about discovering new ones that can add another dimension to our gardens. Each time I come upon a plant that blooms during a low-colour time of year, or one that grows well in difficult situations, I get excited.

Our 11-hectare (27-acre) show garden, situated in zone 6 with extreme heat in summer, very cold winter winds and more rain than I care to mention, has been a good trial ground for many plants mentioned in this book. I also travel a great deal and seeing so many plants, in a wide variety of situations in diverse areas of the world, has really broadened my perspective. The old adage, "the more you learn, the more you realize how little you know," certainly applies to gardeners—at all levels of experience.

I value the friendship of many people from around the world who so willingly shared not only their knowledge of plants, but also their inspirational love for them. These folks have been incredibly helpful in my choice of plants for this book.

Use this book as a springboard for your own ideas and experience. Read, don't be afraid to experiment, and keep a record of plant performance and weather patterns for success with growing. For instance, you'll see zone designations throughout the book, which represent the average minimum temperature to be expected.

You and I know, however, that weather patterns are not consistent and some winters may dip below the average. There are also wide variations within zones, and even within neighbourhoods, depending on wind protection, frost pockets and so on. So use the zone designations as a guideline, but rely on first-hand knowledge of your own weather conditions. (This is where your record-keeping comes in.)

The plants I recommend here are by no means the only top choices. They are simply the ones I have selected as being great plants that have proven themselves. I hope this book will help you make the best choices for your garden.

Zones and temperatures

This chart shows the average annual minimum temperatures that can be expected in each zone, as well as examples of plants that will succeed in each zone.

Zone	Temperature	Common name (botanical name)
1	Below –50°F Below –45.6°C	Lapland rhododendron (*Rhododendron lapponicum*) Netleaf willow (*Salix reticulata*) Pennsylvania cinquefoil (*Potentilla pensylvanica*)
2	–50°F to –40°F –45.6°C to –40°C	American cranberry bush (*Viburnum trilobum*) Bush cinquefoil (*Potentilla fruticosa*)
3	–40°F to –30°F –40°C to –34.5°C	Common juniper (*Juniperus communis*) Winged spindle tree (*Euonymus alata*)
4	–30°F to –20°F –34.5°C to –28.9°C	Vanhoutte spirea (*Spiraea x vanhouttei*) Virginia creeper (*Parthenocissus quinquefolia*)
5	–20°F to –10°F –28.9°C to –23.3°C	Boston ivy (*Parthenocissus tricuspidata*) Flowering dogwood (*Cornus florida*)
6	–10°F to 0°F –23.3°C to –17.8°C	Japanese maple (*Acer palmatum*) Winter creeper (*Euonymus fortunei*)
7	0°F to 10°F –17.8°C to –12.3°C	English holly (*Ilex aquifolium*) Kurume azalea (*Rhododendron* Kurume hybrids)
8	10°F to 20°F –12.3°C to –6.6°C	Strawberry tree (*Arbutus unedo*) *Pernettya macronata*
9	20°F to 30°F –6.6°C to –1.1°C	Fuchsia (*Fuchsia* hybrids)

Getting to the Root of It
Soil and Compost, Seeds and Seedlings

All too often we make the mistake of judging a garden by what's under our noses—the intoxicating fragrance of a rose or the spectacular sight of a rhododendron in full bloom. But I tend to think about gardens first and foremost in terms of what's at my feet—the soil. It's where plants take up nutrients, where water is absorbed, and where life for seeds begins. So before you think about which varieties to plant, spend some time thinking about improving your soil. I promise you won't regret it.

Soil and Compost

Soils aren't simply lifeless dirt; they are a mixture of organic and inorganic material, microorganisms, nutrients, air and moisture. Most soils fall in a range somewhere between clay (heavy) and sand (light). Clay soils tend to hold too much moisture and restrict air circulation. Sandy soils drain too quickly and don't hold nutrients needed by plants. Some plants prefer slightly sandy or slightly heavy soils, but most prefer a middle range of soil, commonly known as loam—a mixture of sand, silt and clay.

Soil types

The first step to establishing your soil's condition is to test its texture. Take a handful of moist earth and squeeze it as hard as you can. You should be able to open up your hand and be left with a ball that crumbles easily when touched. The ideal balance is a soil that is heavy enough to hold some moisture but light enough to drain well and allow air to circulate.

If it won't stick together at all during the squeeze test, you need to add organic material, such as a combination of compost and well-rotted manures. It's best to add these in the fall if possible, but they can be added early in the spring as well.

Most plants will do better in soil supplemented with organic matter, which builds up the bacteria in the soil. Manures will also add some nitrogen, although not enough to balance the addition of a large amount of sawdust or bark mulch.

If your soil is heavy and clumps together, a combination of peat moss, bark mulch, sawdust and sand should set things right. Use one part peat moss, one part bark mulch or sawdust and slightly less than one part sand. Keep adding this to the existing soil until you achieve the proper consistency. If you are adding sawdust or bark mulch, you will also have to add a bit of nitrogen, such as sulphate of ammonia or 21-0-0, because the bacteria that break down sawdust and bark mulch tie up the soil's nitrogen. The addition of nitrogen helps break them down more quickly. By spring the material should be well composted and ready for use in the garden.

If you want to avoid nitrogen-boosting additives, then compost the sawdust or bark mulch for 6 to 12 months before putting it in your garden.

When using sawdust and bark mulch, make sure that it is fir or hemlock and *not* cedar, which contains resins that can be toxic to many plants. If you're adding sand, don't take it from a saltwater beach—the salt is bad news for many plants.

Gypsum, available at garden centres, is also useful for breaking up heavy soils. It binds small particles of clay together to make the soil more crumbly. Worms are another great helper in aerating soil and assisting with the composting process. You can never have too many worms in your garden.

Soil pH

The pH balance of soil is an important factor in determining how well your plants perform. If the pH is off, the plants are often unable to obtain nutrients from the soil. You want to keep it between 5 and 8. Simple pH test kits are available at garden centres.

The other option is to take a soil sample and send it off for a full-blown soil test, which gives a thorough breakdown of the soil sample, including micronutrient information. You can find a laboratory that does soil testing in the Yellow Pages or through a local garden centre.

Lime is sublime

In areas of heavy rainfall, the soil becomes quite acidic over time, sometimes dropping the pH below 4. When the pH is too low, some plants will not grow well because they cannot pick up nutrients that are locked in the soil.

 shot of rye

Having an organic garden is important for many people, and a good way to boost your soil organically is to throw a handful of fall rye seed into your garden plot. It keeps it green over the winter, but the real bonus comes in the spring, when you dig the rye under, releasing nitrogen into the soil. Rye seed cannot be stored for over a year, so buy it fresh each fall to assure good germination. Coverage is about 1 kg per 50 square metre (2.2 pounds per 500 square feet).

If you live in the coastal region, put a bit of dolomite lime into the soil at the time you sow fall rye (except in plots where you will plant potatoes). The lime adds magnesium and makes the soil more alkaline—a plus for most vegetables.

An alternative winter crop is a blend called Rejuvenation Mix. It contains a blend of winter peas, winter wheat and other crops. Peas have nitrogen-fixing bacteria in their root nodules. When the roots are dug into the soil in the spring, that nitrogen is released back into the soil. This mix doesn't look as good as fall rye, but it works more effectively.

To correct a soil that's too acidic, apply lime at the rate of one 20-kg bag per 93 square metres (44-pound bag per 1000 square feet). But don't use lime where you're planning to plant a patch of potatoes (it can lead to problems, such as scab, with your crops). If possible, always apply lime in the fall rather than waiting for spring. That way the lime will release at the same time winter rains are pushing pH downwards and the pH reading won't fluctuate like a roller coaster from one extreme to the other.

Always leave at least seven days between applying lime and manure.

Several types of lime are available. Here's the rundown on the differences.

Ground limestone is the cheap stuff. It works fine, but takes longer than the alternatives. Apply it in the fall and about four months later it will have kicked up the pH.

Hydrated lime is without a doubt the fastest lime, but is it ever messy. It is dehydrated into a white powder that is a pain to apply. You'll have it everywhere—in your shoes and clothes—and if it's windy, it will cover most of the town, too. When water hits it, though, it goes to work instantly. Lots of folks use it to make a white paste they paint on trees. Chances are they're not too clear on why they paint it on the trees, except that their fathers and grandfathers did it. Keeps the bugs out of the trees, they say. Right? Wrong! But it does expose the bugs to their natural predators. Insects' natural camouflage doesn't work on an all-white background, and they get picked off. So while the lime doesn't really deter the bugs, it helps get the job done just the same.

We also use it to whitewash benches in our greenhouses. If you have your own small greenhouse you can use it on glass or plastic as a whitewash too, but mix it with milk for best results.

Rapid lime is faster than dolomite and takes about four weeks to release into the soil. The main advantage is that it's easier to apply. It's powdery and contains magnesium, which will help green your lawn.

Dolomite lime is traditional garden lime and is widely available. It takes six to eight weeks to act, so apply it in the fall. The speed at which it works depends partly on the grind. Bigger granules mean slower release times. Dolomite lime comes in different degrees of "fines." Landscapers often use coarser grinds because it is easier to apply, while greenhouse growers tend to prefer the finer grinds for faster results. Like rapid lime, dolomite also contains magnesium.

Dolopril is a new generation of limes that is "prilled" for ease of application and quick releasing. It is more expensive, but provides better coverage, so the cost works out about the same.

Alkaline soils

In more arid climates, such as B.C.'s Okanagan region and the prairies, the soil tends to be at the other end of the pH scale—the alkaline. To drop this down to where plants are more comfortable, sulphates work well. Use aluminum sulphate in a soluble solution dispensed from a watering can or hose-end sprayer. The other alternative is to use a sulphate fertilizer. The addition of peat moss helps, too.

hodos love acid
Some plants, such as rhododendrons, prefer slightly acidic soil, so be careful when adjusting pH around them. Also, don't use manures on azaleas, rhododendrons or other acid-loving plants, such as heather and camellias.

Fertilizer basics

As well as good soil texture and appropriate pH, plants need three major nutrients and several minor ones to make healthy growth. Most fertilizers are labeled with their three main ingredients—nitrogen, phosphorus and potassium—on the front of the package. But what does 20-16-10 mean compared with 20-20-20? The three numbers represent the ratio and amount of nitrogen, phosphorus and potassium inside. Their chemical symbols are N, P and K.

Nitrogen (N) stimulates leaf and shoot growth. It's the most important ingredient to get your plants growing quickly. If soil is nitrogen deficient, plants may lack vigour and look spindly, and the leaves might have a yellowish hue. An example of a high-nitrogen fertilizer is 21-0-0.

Phosphorus (P) is important for healthy, strong roots, the development of fruit, and disease resistance. In most gardens where fertilizers have been used over several years, phosphorus levels tend to be high rather than low. Superphosphate 0-18-0 is an example of a high-phosphorus fertilizer. Bone meal (3-16-0) is an organic high-phosphorus fertilizer. If you are planting root crops or bulbs, or starting new crops, bone meal will give newly rooted plants a boost. Remember, the finer the grind the quicker acting but shorter lived it will be. Larger grinds equal slower release over a longer period.

Potassium, or potash, (K) plays a key role in photosynthesis and the plant's ability to resist disease, cold and drought. If plants are deficient in potassium, they may have poor fruit production, underdeveloped root systems and streaky or spotted yellow leaves. Sulphate of potash (0-0-50) is an example of a potash fertilizer. Rock phosphate is an organic high potash fertilizer.

Other nutrients, such as calcium, magnesium and sulphur, and the trace elements zinc, manganese, iron, boron, chlorine, copper and molybdenum are found in some fertilizers. They can also be purchased in a mix known as fritted trace elements. In the spring, especially in areas of heavy rainfall, it's a good idea to add a blend of these micronutrients to the soil. Your garden patch could be deficient and adding them is a safe way of eliminating a problem that's sometimes hard to diagnose. The other option is to take soil samples and have them analyzed at a lab, which can tell you if you have a deficiency.

Now you can see why different fertilizers have different formulations. Your soil may be deficient, or your plants may need different nutrients at various times in their growing cycle. For instance, a high-nitrogen fertilizer is good to get your tomato plants off to a good start, but later in the season more potassium is needed to help them set buds and grow fruit.

Compost

By now many of us have a compost of some sort in a corner of the yard. In urban areas, many municipalities provide efficient and easy-to-use composters at a reasonable price. In

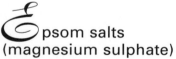

Epsom salts (magnesium sulphate)

In other sections of this book, you might see me recommend Epsom salts to treat a problem. Epsom salts work as a greening and toning agent to correct a deficiency of magnesium. Epsom salts should not be applied at the same time as other micronutrients because they are not compatible. Mix 28 grams (1 ounce) of the salts with 8 litres (2 gallons) of water. It's also great for sore feet!

addition to cutting down material sent to the landfill, composting plays an important role in our gardens, adding nutrients and improving the texture of soils. The organic material in compost is wonderful for your garden. It helps keep moisture in the soil and slowly releases nutrients once it is mixed in. Finished compost *isn't* soil, but it's one of the organic ingredients you need to build and maintain healthy soils. It's best added to your garden before planting a new crop. Work it into the top 30 cm (12 inches) of soil with a fork or spade.

Composting can be done in a heap, a barrel, or right in the garden itself, depending on your garden and your municipal bylaws. For most yards, a barrel-type composter can conveniently handle a moderate volume of kitchen and yard waste. For larger volumes, four strong metal stakes and a large piece of chicken wire will do the job admirably. Rats can be a problem in urban areas and there are regulations governing the type of composter you can use. Check with your local municipality.

Compost needn't (and shouldn't) be smelly or unsightly if you follow two simple rules. First of all, don't hide it away out of the sun in some dark corner. A composter works best where there is at least partial sunshine. The second secret for success is to use the right mix of ingredients—a composter can't live solely on grass clippings. Think layer cake: ideally you want roughly equal layers of green material, such as grass clippings, alternating with soil or other organic materials such as peat moss. They also tend to break down more quickly when blended together.

Most organic matter can be composted—leftover vegetables, leaves, grass clippings—but too much of any one thing can upset the balance needed to maintain the compost. For example, countless bags of green grass clippings in the summer months can gum together and deprive the compost heap of air. Dried fall leaves are perfect for balancing out the green grass clippings, so one solution is to store your leaves in plastic bags and mix them in with the rest of your garden and kitchen waste over the course of the year.

Starting a new compost

If you're starting a new compost heap or bin, begin with a few handfuls of sticks and twigs on the bottom. This will provide room for air to circulate. Then add a mixture of dried leaves or other brown material, such as dried grass clippings. Green material and kitchen waste goes in next. Then toss in a few shovelfuls of garden soil or peat moss. Mix it up and let the microbes do the work.

From then on it's just a matter of adding a balance of green and brown materials to keep the compost going. Cut up larger items and they will decompose quicker. Give it a mix every week or so to keep the air circulating inside. The worms and insects will soon show up to help things along.

If you want to speed the process up, use a compost booster, which adds nitrogen to the mix. I like to add a sprinkling of liquid 21-0-0 to supply the bacteria with the nitrogen they need.

Raised beds

Raising the soil level of a bed above its surroundings benefits your plants by keeping them warmer in the spring when they need it and assisting with drainage. There is a drawback, though—water leaches out nutrients faster from raised beds than from conventional garden plots. After heavy winter rains, I recommend adding some micronutrients to the beds in the spring. In addition, make sure you toss in a healthy heap of good old manure.

Compost should be the texture of a wrung-out sponge—moist but not damp. Most compost barrels should be covered, but they may need additional water from time to time. The moisture creates an environment that stimulates bacterial growth. Too much moisture and your mix becomes anaerobic and begins to smell bad. A working compost doesn't stink.

If you're making compost in a heap exposed to the elements, make sure it has a well-rounded dome on top so it sheds water and snow in the winter. A tarp over the top is a good idea in particularly wet areas. In the summer, shape the top into a depression so it will hold the water. Remember to check local bylaws before constructing any compost bin or heap.

To speed up decomposition, make sure those microbes have adequate ventilation. That means your composter should have plenty of air vents and you should stir up the mix every week or so. You can also use a stick to poke vertical "ventilation shafts" down into the middle of the mix. Compost works faster when it has some heat to get it working, so place yours in a sunny location. Once it is established, a compost will generate its own heat through the decomposition process. A properly working compost should be hot enough to kill the seed of most weeds.

It's possible to compost even if you don't have a yard. Worm composting is growing more popular these days and can be done on a balcony or under the kitchen sink in a worm box. The worms rapidly break down kitchen waste into rich loam. Some municipal governments supply worm composting starter kits and information for a small fee.

What's what in the compost heap

In addition to yard and kitchen waste, there are other things that can help your compost along. And there are some things you should avoid because they're bad news.

Seaweed, which contains a lot of micronutrients, is a wonderful addition to your garden or compost pile. Layer it with other material in the compost pile to avoid having a gooey mess, and don't layer it more than 5 or 8 cm (2 or 3 inches) deep in your garden at one time.

Some people worry about adding weeds to the compost for fear of reinfesting the areas where the compost is used. As long as you're composting correctly, it should be sufficiently hot to kill the seeds. A temperature of 180°C (355°F) will kill weed seeds, and believe it or not, most composts will generate that much heat on their own in the summer. But why take a chance? To be safe, I don't recommend adding the mature flowers of weeds to the compost.

Good candidates for the compost pile:
- lint from the dryer (don't add a huge amount all at once)
- most fruit and vegetable matter, as long as it has no added oil, etc.
- coffee and tea grounds
- grass clippings (uncontaminated by weed killers)—no more than 8 cm (3 inches) deep at a time, layered with other material
- leaves
- sawdust and fine bark mulch, except from cedar and walnut trees
- chicken, cattle and horse manure
- shredded newspaper (most use organic inks these days)
- straw
- seaweed—no more than 8 cm (3 inches) deep at a time, layered with other material

---

More information

If you have a question about composting, contact your local government. They are eager to help gardeners with composting tips, equipment and demonstrations.

- soil
- peat moss
- chipped or finely cut up branches and twigs from pruning

Here's what to avoid:
- diseased plants
- meat or dairy products
- rhubarb leaves or walnut (they can be toxic or phytotoxic)
- sawdust from cedar or walnut trees
- loads of evergreen needles (they increase the acidity)
- anything treated with weed killer, including grass clippings
- pet or human waste

Manure—gardener's gold

Manure is animal waste that can be worked back into the soil to boost nutrient levels and organic content. Whenever I talk about adding manure to the garden it's always prefaced by the phrase "well-rotted." By that, I mean manure that is several months old, so that bacteria has started to break it down. This makes the nutrients more readily available to the plants and prevents damage—fresh manure can burn tender plants.

If you're starting with the fresh stuff, begin composting it in the fall, and it should be ready for your garden the following spring. The test is that it should crumble easily. If you're buying packaged manure from a reputable source it should be well composted by the time you get your hands on it.

Some manures are safer than others. When you're dealing with animal waste, there is always a slight danger of harmful bacterial contamination, and that's why it's better to go with the critters you know. Cows, bulls, calves, heifers and steers are all bovines, just with slightly different configurations. (City slickers will simply have to take my word on this one.) Horse, chicken, sheep and swine are the other readily available sources of manure. Mushroom compost is second-hand horse manure that's been used for growing mushrooms. It has lime added, making it good stuff. It also is best if allowed to decompose for another six months.

Remember, well-rotted manures are a great organic addition to our gardens, but they tend to be a bit alkaline. For this reason, they should not be used on rhododendrons, azaleas or any broadleaf plants, conifers and heathers. Manure can cause scab on potatoes, so keep it clear of the potato patch.

Compost vs. manure

Compost and well-rotted manure are both integral ingredients in productive soil, but which is better? Compost puts organic matter back into the soil, but manure contains more nutrients. In other words, you should be adding both to your garden.

But don't make the mistake of expecting compost and manure to work as soil. It's good stuff but it's not soil! A good soil is the sum of its hardworking parts. To get there, I can't emphasize enough the importance of adding organic materials such as manures and compost, adding micronutrients, and planting fall cover crops to help build up the soil.

Mulch

Mulch is marvelous material that makes life easier for the average gardener. Surprisingly, many people don't use mulch despite its benefits. Gardeners use organic or inorganic material for mulch, but I prefer organics where possible.

Organic mulches

Organic mulch creates a physical barrier that reduces surface evaporation and keeps new weeds from sprouting. Mulch also insulates plant roots from temperature fluctuations. In summer, a layer of mulch cools the soil, promoting root growth.

Here are some suggestions:

- grass clippings
- peat moss
- pine cones/needles
- leaves
- bark chips
- compost
- wood chips/shavings
- newspaper
- straw
- coconut fibre

Note that newspaper works as a mulch, but it won't last as long. And it's not particularly attractive, so you might want to cover it with other material. If mulching with straw, make sure it does not contain seed tassels, which will germinate and leave you with a whole new crop of weeds to deal with.

Inorganic mulches

Inorganic mulches are best used over a layer of landscape fabric, which allows moisture to penetrate but reduces weed growth. Put down a 5- or 8-cm (2- or 3-inch) layer of mulch around your plants, making sure to use enough to hide the landscape fabric. There are a number of different grades, but I recommend avoiding the cheaper ones, which often deteriorate rapidly. The commercial grades will save you time and money in the long run.

- aggregates (gravel, shale)
- rocks
- bricks
- lava rock

Get digging!

Conditioning your soil is an ongoing project, but late February is the time in milder areas to get the soil ready for spring planting. Those in colder climates have to wait a little longer. In coastal areas, winter rains beat the soil down and make it more acidic, so this is the time to check the pH and adjust it with some lime if you didn't do so in the fall.

The second task is to put organic matter into the soil. This is where the compost and well-rotted manures come into play. Get that shovel ready!

The third step is to raise the soil level by berming it up. Using a rake, hill it up at least 20 cm (8 inches) high. The plants will be higher, drier and much warmer and you will get them off to an early start. This simple step can raise soil temperatures by several degrees.

Soil timetable

Spring. Turn over soils beaten down by rain. Check pH. Dig beds deep and turn over. Add organic matter and whatever else it takes to break up heavy soils. Remember the squeeze test!

Summer. Use mulch to protect plants and conserve water. Fertilize as required.

Fall. Dig in mulch. Add organics such as compost and bark mulch to break down over the winter. Plant fall rye or Rejuvenation Mix in vegetable gardens.

Winter. Continue to add organics. Turn over periodically.

These are the basics when it comes to preparing your garden soil. Whether you're planting flowers or growing cucumbers, following these guidelines should help you get your yard ready for planting. Healthy soil really is a key to gardening success—it does everything from assisting the growth of plants to combating pest infestations. Spend time preparing the soil and you will see the difference in many ways.

Seeds and Seedlings

Growing from seed

Sometimes great things can come from scattering a few seeds in the garden and scratching them into the earth. Other times, you might need a greenhouse and an encyclopedia's worth of knowledge to achieve the desired result. Level of expertise, money and time constraints are just some of the things that come into play when deciding whether or not to grow from seed. I take great satisfaction in the hands-on approach of starting from scratch, but you might be happier simply buying bedding plants and popping them into place. There's nothing wrong with either approach, and it's up to you to decide what you're comfortable with.

When buying seeds, try to resist the temptation to go wild when you're in the garden shop or filling out the order form for the seed catalogue. Take a walk around your garden with a notebook and plan what you want to grow in each location. Don't forget to bring the list with you when you go to the store.

Ordering by mail is also a lot of fun. Seed catalogues are a wonderful resource, are free, and contain a wealth of information. Order a couple in the fall for terrific winter reading and garden planning. You will generally find vast selections of both hybrid and nonhybrid varieties, and many unusual seeds. Unless you have a huge garden, the trade packets are usually the best size to order, rather than the "per ounce" price. The colour pictures look so inviting on those cold winter nights, it's easy to get carried away. I've been guilty of this in the past, so now I have hard and fast rules.
1. Write out my wish list of seeds along with their prices on a big piece of paper.
2. Go away and do something else for a day or so.

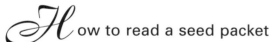ow to read a seed packet

Seed packets contain a lot of information in a small space, so keep your reading glasses handy. Some of it might seem a bit cryptic, but if you read carefully you can prevent unexpected problems.
- Select seed varieties that will grow in your climate zone and weather. Just because you can buy the seed doesn't mean it will grow in your back yard. Buy seed from someone who can give you good advice.
- Some seeds are easier to start than others, so know what you're getting into. Determine if you can sow them in the garden or if you'll have to start them indoors.
- Is it difficult to germinate or does it make more sense to purchase starter plants?
- Find out how many days from planting to maturity. Is there enough time for the plants to mature before frost hits in the fall?
- Look at the soil type needed. Should it be rich and full of humus, or sandy loam?

3. Come back and look at the prices. It may be only $2.95 here and $1.95 there, but it adds up fast!
4. Ask myself if I have room and time to plant everything on the list. (My wife wonders about this when she hears me arguing with myself.)
5. Rewrite the list and finally place the order. Despite the rules, I still end up ordering too many seeds!

Planting seeds directly in the garden

Ever run into someone who dumped a few packets of seed into the back garden plot and wondered why nothing ever came of it? It happens all too often and is often the result of a few avoidable mistakes. Here are a few tips to make sure at least something pokes its head out of the soil.

- Prepare the soil well. See the sections earlier in this chapter to make sure your soil is ready to host the seeds.
- Raise the beds as much as possible. This gives better drainage and warmth to your plants. Raising the beds even slightly above the level of the surrounding ground can make a big difference in warmth and moisture.
- Rake the top layer smooth, remove any rocks or large debris and add a fine layer of peat moss or sterilized potting soil as extra help for young roots.
- Think about the final look—plant height, colour, and light requirements. If you're sowing a vegetable plot, plant in rectangular blocks rather than rows; it makes better use of space.
- Most people plant seeds too deeply, but it's cold and wet down there! Keep them higher and warmer. It'll make a big difference in germination. This is where raised beds help give seeds a boost with their warmth and better drainage. Remember, you can raise the beds simply by mounding your soil higher than the surrounding area.

Harvesting seed

Seeds don't all come in packets. If you want to harvest your seeds (or your neighbour's) for next year, follow these steps. Do remember, though, that plants grown from seed you've collected may be different from their parents and may not have the special characteristics of the parent, such as height or colour variegation.

- Let the seeds mature on the stem. This is very important. They should almost be falling out of the pod.
- Put them on a paper towel and allow them to dry in an area with good air circulation.
- When they are thoroughly dried, package them in envelopes, label, and store in the refrigerator or freezer over the winter. This stratifies them—a horticultural term that means giving them a period of cold, which aids germination. I use the freezer, but many experts recommend the refrigerator. Whichever you choose, you'll know where they are when it comes time to plant.
- Most seeds can be stored for years. If you're just freezing them to stratify them, 48 hours is the minimum time for most seeds.

- Keep the seeds moist until they germinate.
- Once they've sprouted, thin them according to seed packet directions.

Germinating seeds indoors

The growing season is too short for some plants to complete their life cycles outside. These plants must be started inside and moved out when the weather is right. Timing is crucial: the plant must be germinated at the right time so that it's ready to plant out when the weather is appropriate. If you plan on having something in the ground May 24, refer to catalogue information to determine the number of days to germination, and the time required to grow the seedling and acclimatize it. Once you have the total, work backwards from the day you plan on having it in the ground.

The key to starting seeds indoors is to keep them consistently at the recommended germination temperature and the right amount of light, and to keep humidity high.

- Stratify the seeds by placing them in the freezer or refrigerator for at least 48 hours. Stratification is a horticultural term for exposing seeds to chilling temperatures. Many seeds would, in nature, go through a period of cold dormancy before blooming, and placing seeds in the freezer mimics this process. Some seeds don't need it, but it can't hurt. Some seeds require up to two weeks in the freezer or refrigerator, so check planting instructions carefully. Leftover seeds can go back in the freezer until next year.
- Fill trays with a seed-starting mix, such as mica peat.
- With a pencil, make a furrow in the soil surface, place the seeds in the furrow and water them into place. Most watering cans are too hard on seeds and end up washing your seeds out. You can get special watering cans with a fine "rose" (the Hawes is the best), or use misting bottles if you don't have too many seedlings.
- Use very warm water to get the seeds started.
- Cover the trays with a plastic cover or polyethylene to keep them nice and humid.
- Many seeds, such as impatiens and begonias, require at least 14 hours of light for germination. I like to leave my overhead lights on 24 hours a day. A single or double fluorescent "cool white" is sufficient. You need light *intensity*, not necessarily grow lights.
- For seeds that require total darkness to germinate, lay some newspaper on top of the seed trays and put plastic on top of that. This will keep the tray dark and humid until the seeds sprout.
- Keep seeds at a consistent temperature when they are germinating. A bottom temperature of 18° to 24°C (65° to 75°F) is ideal. Special heaters are available for this purpose, but are not a necessity if you have the right room temperature.
- Once they germinate, remove the covers. Move the lights closer to the seedlings so they don't get leggy.
- To prevent disease, don't overwater and make sure you have good air circulation. A household fan is great for this purpose. Use fungicides to prevent damping off and other diseases that cause seedlings to rot or wilt.
- Once the second and third set of leaves begin to develop and the outside world is frost free, move the seedlings into a cold frame to ease them into the garden. Some seedlings will turn purple if it's too cold. If this happens, move them to a warmer spot until they are green again.

Toughen seedlings in a cold frame

Whether you've bought them or grown them yourself, your seedlings take the same path into the garden. For best results, don't just plunk them down in the soil now that you've come this far. There is an intermediate step where you have to ease the little guys out into the cold world.

Your seedlings are going from a climate-controlled setting to the variable weather outside and they must be acclimatized over several days. The easiest way to do this is in a cold frame.

Cold frames come in a variety of shapes and sizes, but are essentially mini-greenhouses that can be easily opened and closed. You can buy them premade or make one yourself from basic building materials and polyethylene. I make mine from softwood and screw the parts together so it can be easily dismantled and stored when not in use. If you don't have a cold frame, cover plants for the night with clear plastic over a simple wire frame.

Basic cold frame

Most plants require seven to ten days before they are acclimatized to outside conditions.

Transplanting into the garden

Transplanting seedlings into your garden is an important step in the growing process. As soon as the first true leaves appear and the seedlings are strong-stemmed and properly acclimatized, they can be transplanted. The best time to transplant is on an overcast or drizzly day, as sun can scorch the delicate new leaves. Reemay cloth or a similar covering will give the plants protection and get them going faster.

- Assuming your planting beds are already enriched with plenty of organic matter, have good drainage and nice texture, spread a layer of good old reliable 6-8-6 fertilizer or bone meal to stay organic over the area to be planted, and rake it in.
- For most plants, loosen the soil to a depth of 30 cm (12 inches) to allow the roots to penetrate easily.
- Dig a hole slightly deeper than you will need to receive your transplant.
- Make sure the plant is thoroughly moistened. Place one hand on top of the container with the stem protruding between your fingers and tip the container upside down, allowing the plant to fall gently into your hand.
- If the roots are tightly matted, gently spread the lowest third of the roots with your fingers.
- Pop the root ball into the soil at the same level it was in the pot. Sometimes tall, leggy plants can be set in a little deeper for extra support. But be careful not to plant too deeply (tomatoes are an exception to this rule; plant them deeply and they will root all along the buried stem).

A mud slurry

When you're transplanting, try to keep the soil around the roots, but if you lose it, or are planting bare root plants, use a mud slurry. To make a slurry, place a few handfuls of soil in a pail and mix with water to form a thick slurry. Add some liquid root starter or fish fertilizer. Dip the roots of your transplants into the slurry, and then plant as above. When you're finished, pour the leftover slurry around the base of your plants to give them an extra boost.

- Gently fill in around the plant and finish up by giving it a thorough watering.
- Keep an eye on your transplants to make sure they're adapting well to their new environment.

If your plants are in fibre pots, do not plant them in the pots. Break the edges away gently before planting them. Be careful not to disturb the delicate roots when you do this. Before setting out plants, dip them in a mud slurry (see sidebar above). This will get them off to a faster start by encouraging the rapid formation of new roots.

Fertilize as soon as the plants become established and you can see growth beginning. A 20-20-20 fertilizer is the most popular, but if you are having some difficulty with the plants becoming spindly, use a low-nitrogen/high-potash fertilizer (0-12-12, for example). Water first, then fertilize. Never feed anything when it is dry for fear of burning. Weekly fertilizing is fine if you want to push things along.

Making the Most of Your Vegetable Patch

Time and space are two things that seem to be in short supply for many gardeners these days. With all kinds of demands on our time and smaller outdoor spaces, many people write off the idea of putting food on the table from their own yard. At the same time it's hard to ignore the value of growing your own fruits and vegetables. The bottom line is a good place to start—people pay high prices for quality produce at specialty stores, but you can grow even better quality produce for a fraction of the cost. A packet of lettuce seed costs about the same as one head of lettuce and will provide greens for your table all summer long.

Today's refined palates crave the exotic and unusual varieties that aren't produced on the mass market; you can add a lot of pizzazz to your meals by growing new and tasty varieties yourself. And home-grown vegetables don't just look good on the plate—today's vegetable patch can be as attractive as any flower bed. And if that's not enough, there is a lot of concern over agricultural practices and the use of pesticides and herbicides. If it comes from your garden, you know exactly what you're feeding your family. With those thoughts in mind, it's hard to imagine why more people aren't growing their own vegetables.

But let's go beyond the basic lettuce, tomatoes and a few rows of carrots. A vegetable garden doesn't have to be a visually boring patch tucked away out of sight—it can be an integrated, interesting part of the landscape. It just takes a little creativity. Garden beds don't have to be rectangular and boring, they can have interesting, curvy shapes. Many vegetables are attractive in their own right. The line between ornamental and edible is blurring all the time—many people plant ornamental kale or cabbages for their wonderful winter colour, not for culinary reasons.

Planning and Caring for Your Vegetable Patch

With smaller gardens, it's important to realize you can't grow everything, so choose wisely. Grow your favourites, but consider those that are the most expensive in the stores. Corn, for example, takes up a lot of space in your garden, and is relatively cheap when it is in season locally. Growing red peppers, tomatoes and cucumbers makes more sense because they are expensive to buy. If you like to freeze or preserve vegetables, then grow pole beans, "telephone" peas and beets by the bushel. With a little cool storage space you can keep onions, winter squash and most root crops well into winter.

In many growing zones (6 and higher), the vegetable patch is an ongoing concern, producing for your table throughout the year, rather than just in the summer and fall. As one crop finishes, another can be started, keeping your garden continually productive. Despite the initial investment, vegetable gardening really does pay off financially.

Whatever the size of your garden, avoid wasting space. We often make wide paths for rototillers and narrow strips for food. Reverse that. Plant in blocks or wide rows instead of narrow strips. You'll have more growing area than pathways, and less space for weeds to take over as well. This is especially important for the urban garden where space is at a premium. Carefully plan the location of your crops so they don't shade other vegetables when they are mature. You'll find that the outside of the rows mature earlier than the shaded and less stressed insides. Work from the outside in for a prolonged harvest.

Preparing the soil

It's possible to take a patch of lawn, remove the sod, amend the soil and get planting in short order. However, if you're planning a few months ahead you can do a better job.

If it's possible, choose an open, sunny spot away from the shade of tall buildings and trees. Most vegetables need a minimum of six hours of sunshine a day, especially between 11 a.m.

*W*atering the vegetable patch

Water in the morning rather than late in the day. Plants absorb moisture more efficiently in the period of time just as the sun rises. Water the soil, rather than the whole plant. Putting water on the foliage can lead to fungus problems such as blight or mildew, which have ruined many crops over the years. Drip irrigation systems are great because they put water onto the plants at ground level, making good use of water and keeping foliage dry. Water thoroughly and deeply to encourage roots to reach down for the moisture. Watering too frequently creates shallow, more fragile root systems.

When it is hot people tend to overwater. But you can prepare your plants early in the growing season for the hot months ahead by encouraging them to send roots deep into the ground. Water once a week to a depth of at least 10 to 15 cm (4 to 6 inches) from the start of the growing season. If you do this religiously, plants will develop root systems deep enough to protect them from drought when the hot weather hits in June or July.

Broadleaf plants exposed to the sun can lose a lot of water through evaporation. Spray the foliage to remoisturize the leaves, but do not water the leaves when the sun is at the burning stage, between 11 a.m. and 3 p.m.

and 3 p.m. Good air flow is also important to help prevent and control fungal diseases.

If you are planning a full year ahead for a new garden site, turn over the soil and throw in a mix of winter peas and clover. They are wonderful for building up nitrogen in the soil. Add some organic matter, such as manure and compost, along with the peas and clover. By the end of the season you will have made a difference to your soil. Don't let your existing garden stay bare over winter—plant some fall rye or Rejuvenation Mix (winter wheat and winter peas) to help protect and enrich the soil. For more tips on preparing soil for a garden, see chapter 1.

Using raised beds

Raised beds are very popular for growing vegetables and with good reason. In wet conditions raised beds give better drainage. They also create warmer soil, which gives plants a boost early in the season. As well, if you have trouble bending down and working at ground level, you can build raised beds or boxes up to a level that is comfortable for you.

Railway ties are popular for creating raised beds, but make sure you know what you're using. Actual railway ties are preserved in creosote, which may leach into your garden beds. Wood sold as "landscape ties" in garden centres or lumber yards is often just dimension lumber which has been treated with a preservative. Most of these preservatives won't leach into the soil, but to be safe you can line the inside of your beds with plastic to be absolutely sure nothing leaches from the ties.

I prefer to use bricks or interlocking blocks rather than wooden ties. As the soil expands in volume from the annual addition of organic material, you can add another layer of blocks to raise the beds. Bricks or blocks look nice in the garden and you don't have to worry about leaching. The bricks also allow you to overcome something that drives me nuts—square beds! With the bricks you can create an attractive shape that blends in with your landscape. Outlined by smooth-flowing, interesting curves, your vegetable garden can include perennial vegetables such as asparagus, and small fruits like strawberries, and perhaps it can be banked on the north side with some vines—grapes, kiwi or thornless blackberries. A vegetable garden can be attractive as well as functional.

Growing in containers

Container gardening has become increasingly popular over the last few years, and fruits and vegetables can often be found popping up in spaces where flowers ruled in the past. Plant breeders are responding to this demand by creating more container-friendly varieties. This is good news for people who want to harvest fruits and vegetables from their balcony containers.

Theoretically you can grow almost anything you want in a container, but consider economics and aesthetics when deciding what to plant. Grow your favourites, the ones that are most expensive to buy in the stores, and those best suited to container growing.

Cucumbers, for example, can set you back a couple of dollars each at the produce stand. Varieties such as 'Salad Bush' are great for containers, with vines that grow about 60 cm (24 inches). You'll have an abundance of 20-cm (8-inch) fruit in about 58 days. 'Bush Pickle', 'Space Master' and 'Pot Luck' are also good bets.

Tomatoes are the number-one choice for home gardeners and there is a wide selection of superb container varieties on the market. 'Patio Prize' is a vast improvement over the old 'Patio' tomato, both in flavour and plant vigour. 'Lunch Box' from Stokes is a good one as well; it's a small tomato with really sweet fruit. If you want something unique in cherry tomatoes, try the new 'Red Robin' or the flavourful yellow 'Cherry Gold'. It's from

England and can be used as a tumbler from hanging baskets.

Peppers do well in containers and look very attractive to boot. An All-American award-winner, 'Super Chili' is a good bet. If you like them hot, this one puts 'Jalapeño' to shame and looks great in containers. My all-time favourite is the sweet yellow hybrid, 'Gypsy'. Remember to keep them in full sun, and don't let them dry out or they will go backwards fast.

If you love the sweet taste of carrots, try award-winning 'Thumbelina', a little, round orange ball that is sweet and ready to eat in a couple of months.

Leaf lettuce is ideal for containers, and a mix of types can provide a display rivaling annuals and perennials. The red-leaf varieties make a great contrast with flowers. If you like head lettuce, 'Mini Green' is the one for you. It is really restrained as far as bolting is concerned, and tastes good even when it's been under stress.

Beans and peas can also work well in containers. Use a window box and tack on a trellis so the vines will have something to hold onto. Scarlet runner beans look good all summer with those gorgeous red flowers and produce tasty beans as well. Peas have to be planted early in the year because they don't like the heat. The sugar snap types are most productive in small spaces—they don't go hard and bitter even under stress, and you can eat the pods.

There are bush squash that actually produce in containers, but use summer squash, which produce crops ready for salads or stir-fries in about 55 days. Yellow summer squash 'Goldbar', scallopini squash and bush zucchini are among the best. (Remember to restrain yourself to just one of those prolific zucchini plants per block, please.)

Everbearing strawberries planted in May will produce loads of delicious fruit by July and a continuous array of colourful white blos-

ips for success with containers

- Use as large a container as possible: 25 litres (7 gallons) is the minimum for tomatoes, squash and cucumbers. You can grow lettuce in a 25-cm (10-inch) pot.
- Use sterilized soil as a base and add composted organics (manure and compost) for valuable nutrients. Make sure the soil is well drained and the container has drain holes. Don't use garden soil in containers.
- Select vegetable varieties suitable for containers.
- If you want to have a carefree container garden, set up a drip irrigation system connected to a timer.

- Plants in containers require more frequent applications of nutrients. Whether you choose to stay organic or use a mix of both organic and chemical, you should fertilize at least weekly during the active growing season.
- Be creative and make those containers look great. Use towers and trellises for climbing vegetables and colourful greens for the edges.
- Put casters underneath heavy containers so they can be moved to take advantage of sun, or shifted out of the way if you need space.

soms as well. I recommend the 'Quinault' variety. All you need is four plants in a 25-cm (10-inch) basket or balcony planter. They will tolerate most winters in climate zone 6 and above.

Growing vegetables year-round

If your garden is in climate zone 6 or above, think of your vegetable garden as a year-round project. Even in midwinter, when growth is slowed right down, there should be a few crops ticking away under that blanket of snow. A good time to prepare for fall and winter crops is late July and early August. Brussels sprouts, kale, parsnips, turnips (rutabagas) and leeks will provide a wonderful supply of fresh vegetables for Thanksgiving and Christmas dinners.

Winter vegetables are basically the same as those grown in spring, summer and fall. The difference is that these varieties grow vegetatively in hot weather and mature in cool weather. The flavour of many of these late vegetables improves dramatically with cooler temperatures, even a frost. True winter vegetable varieties, available from several seed companies, go in the ground in late July, or as transplants by mid-August, for harvesting the following January through March. If you've never tasted your own fresh cauliflower, cabbage or winter lettuce in January or February, you've missed a real treat. And don't forget the perennial and biennial herbs — mint, parsley, chives, marjoram and oregano, which you can harvest through the winter.

The spinachlike greens of Swiss chard are a necessary part of every garden, as far as I'm concerned. The multicoloured foliage and stems of 'Bright Lights' Swiss chard is also a great way to add colour. Once established, it will tolerate a fair bit of cold before it freezes or goes dormant. The same is true of spinach. Cold-resistant 'Savoy' (Stokes) or 'Bloomsdale Savoy' (Territorial Seeds) are two varieties ready for harvest after 45 days.

Late summer is not too late to plant root crops, such as beets. In the worst-case scenario, you will only get the greens, but these are delicious in their own right. Novelty varieties, such as 'Little Egypt', mature in just 34 days. Carrots can be planted in August as well. Miniature varieties, such as 'Baby Finger' and 'Minicor', mature in about 50 days, while many of the early varieties, such as 'Nantes', take only a few days longer.

Almost any type of lettuce is suitable for midsummer planting and fall harvest. The buttercrunch types mature in about 60 days, while the looseleaf types, such as 'Black Seeded Simpson', 'Super Prize', and 'Red Sails', will be ready in about 45 days. If you prefer slower-maturing head varieties, plant 'Frosty'. It will be out in your garden, ready to pick, in October, when lettuce is getting expensive at the supermarket. The very hardy 'Corn Salad' is a miniature salad green that can be harvested in 45 days, and a frost only enhances its flavour.

Some people purchase onion sets, store them in a cool, dry place, and keep planting them every six weeks right through the summer. Of course, the onions get smaller as time goes by, but you've still got the greens. Some varieties of bunching onions can be planted for fall harvest because they mature in 60 days. Up until that time, you have the greens to enjoy. There are also winter-hardy varieties you can plant in the fall for harvest in the spring: 'Kincho' and 'Hardy White Bunching' will survive most winters, even in cold interior climates.

Winter cauliflower, such as 'Armado Spring', and cabbage, such as 'January King', will mature over the winter for harvest in February. 'Purple Sprouting Broccoli' is another winter favourite. Many gardeners put in some 'Siberian' and 'Westland Winter' kale in late summer. To round things out, winter-hardy leeks can go in the ground.

Research done at the Canada Research Station in Agassiz, B.C. (zone 6), indicates that during one winter in six, on average, the temperatures will drop to the point where you may lose your more tender winter crops. By locating them in more protected areas, you can improve those odds.

It's really quite an exhaustive list, and you probably won't have enough room to try them all. The good news is that you don't have to look at a bare patch all winter long after the tomatoes, beans and peas die. There's plenty of fresh taste to look forward to from your winter garden.

When to plant

If you live in a moderate climate (zone 6 or above), your gardening year can begin in January, by brushing the snow aside and harvesting hardy herbs such as chives and parsley, or winter vegetables like corn salad, lettuce or endive. Cold- and frost-tolerant vegetables, such as broad beans, peas and onions, can be set out while light frosts are still being experienced. Main crop vegetables like early brassicas, radishes, and Swiss chard should be planted when all danger of frost is past.

But don't go crazy too early! At the first sign of spring, some folks plant everything on their list. A week later, a cold snap hits and most of their crop bites the dust. Know which plants are sensitive to cold, and which ones are hardier. Heat-lovers—tomato, pepper, cucumber and squash—should only be planted when the weather warms up to an average night-time temperature of 10°C (50°F). As a general rule, the May long weekend is the time for planting tender varieties in coastal areas. In colder zones, plant when all danger of frost is past.

Seed catalogues are a good guide for seeding times in your area, but keep notes to adjust dates for subsequent years. Remember, too, that weather patterns vary from year to year, so a little guesswork is necessary. My experience has proven over and over again that, when in doubt, plant a little later rather than earlier. You'll never be wrong that way.

Seeds or seedlings

Whether you grow vegetables directly from seed or start out with seedlings depends on a few factors. Some plants, such as carrots, are best sown right in the garden while others, such as tomatoes, should be transplanted. If you can grow from seed, you might ask, why bother with transplants at all? In a word, speed. Why wait a few weeks for seeds to sprout and grow outside when you can get a jump on growing and enjoy those vegetables while the neighbours are still buying theirs at the supermarket?

Seed racks in local garden stores usually contain tried and proven varieties that perform well in your region. The selections in those racks has improved over the years to meet the demand for more variety. Old-fashioned varieties are available because of their popularity and lower cost, but new hybrids are also finding their way onto seed racks. Hybrids often have distinct advantages over the old varieties, and in the case of tomatoes, for example, those would be my first choice.

As with any seed buying, don't lose your head when you're in the garden shop or filling out the order form in the seed catalogue. Draw a scale diagram of your plot and estimate how much room you have for each vegetable you want to grow.

I recommend planting the following vegetables as seedlings: brassicas, celeriac, celery, cucumbers, eggplant, leeks, head lettuce, melons, Spanish onions, peppers, pumpkins, squash, and tomatoes. Most others can be sown directly in the garden.

You can simply buy seedlings from the garden centre, or you can start them yourself in trays or pots and grow them in a greenhouse or under grow lights. You might want to try both ways and see which gives you more enjoyment. If you're new to gardening, start out with the easy-to grow varieties and work on the tricky stuff down the road.

If you're buying seedlings, make sure they are healthy plants that haven't been sitting on a shelf too long. Inspect them carefully, looking for healthy foliage, strong stems and good colour. Don't buy them if they're purple, hard and tough-looking. They should be a rich, dark green and lush. This is especially true with brassicas.

When we're looking at tiny seeds or seedlings it's easy to forget just how much production can come out of one plant. Most of us plant too much and end up trying to figure out what to do with all that extra zucchini. You've been warned—plant according to your needs or come up with a plan to make sure the food you grow is utilized. Some food banks will happily take your fresh produce.

See the "Seeds and Seedlings" section of chapter 1 for information on growing from seed and transplanting seedlings.

A Vegetable Sampler

I have chosen these particular veggies because they are so popular, and in some cases there are so many varieties of each it's confusing to sort out which are best. Keep in mind that other varieties may perform better for you than those recommended—ask around about some local favourites. The ones I mention are tried and true winners in a wide range of climate conditions, and I've even sneaked in a few of my personal favourites. For specific information on seeding and growing each vegetable, consult the seed packets and growers' catalogues—a wonderful source of information.

Asparagus

Asparagus has been cultivated for a long time and is one of our most highly prized perennial vegetables. The name is derived from the Greek word *asparagos,* meaning "to tear," which relates to the prickly nature of some of the stalks.

Asparagus is a relatively expensive vegetable because it takes at least three years to bring into production from seed. It also requires considerable commercial growing area, and it has a limited production season. If you really enjoy this unique vegetable, why not grow your own? Hardy to zone 3, asparagus will tolerate temperatures as cold as −40°C (−40°F). To keep costs down, commercial growers usually start asparagus from seed; most home gardeners start with two-year-old roots. Better yet, you can occasionally find jumbo four-year-old roots in some nurseries, thereby reducing maturity time. Once established, asparagus roots are productive for at least 15 years. 'Mary Washington' has traditionally been the favourite variety, but newer all-male hybrids, like 'Jersey Knight', actually produce larger crops of big, attractive green spears with purple bracts and tight purple tips.

To grow asparagus, you need full sun and well-drained, slightly alkaline soil. It is important that asparagus roots go straight down, so the traditional method of planting involves trenches. Dig furrows or trenches about 30 cm (12 inches) wide and 30 to 45 cm (12 to 18 inches) deep, depending upon the length of the roots. Rows should be 1.2 m (4 feet) apart. Fill the bottom of the trench with 5 to 8 cm (2 to 3 inches) of composted cow, poultry or mushroom manure. Sprinkle a good rooting fertilizer, like 4-10-10, on top at the rate of .9 kg per 2.3 square metres (2 pounds per 25 square feet). To prevent burning, it is important to keep the fertilizer and manure away from the roots, so stir this mixture up well with the existing soil, then add a few inches of soil on top. Create a mound of soil in the centre of the trench, leaving the crest about 10 cm (4 inches) below the level of the garden soil.

At this point, the asparagus roots can be planted. To speed up the rooting process, I always dip them in a mixture of warm water, liquid fertilizer and soil, called a slurry. This murky concoction sticks nicely to the roots, and the fertilizer will immediately begin to stimulate root development.

Lay the roots on top of the mound of soil in the trench, spreading them evenly on both sides. Place the plants about 45 cm (18 inches) apart and backfill the trench, leaving the crowns, or tips, of the asparagus just barely covered with soil. Root growth will begin almost immediately.

Weeds can be a problem. In new beds, well-established weed roots will intermingle with the asparagus roots. To keep the beds weed free and to avoid injuring any asparagus roots, practice shallow hand cultivation.

To keep asparagus active and growing during the summer, the roots need deep watering. During dry spells, soaker hoses left running for several hours at least once a week are the best means of thoroughly watering asparagus beds. You want actively growing shoots for food production and also to develop strength in the roots. As the plumes begin to develop, feed the plants with a high-nitrogen fertilizer. For long-lasting results, I prefer to use a slow-release lawn food like 16-4-4.

If you grow organically, parsley planted with asparagus gives added vigour to both. Planting tomatoes near asparagus deters asparagus beetle because of a substance called solanine in tomato plants. However, if this does not work, you can dust with organic rotenone after the harvesting period.

The second year after planting, you can begin harvesting a few spears for a period of four to six weeks. When they are 15 to 20 cm (6 to 8 inches) high, cut the spears at a 45° angle about 4 cm (1½ inches) below the soil line. Be careful not to damage the crowns. At the start of asparagus season, you can probably gather spears every three days, but as the soil becomes warmer, let them develop into foliage. The appearance of very thin spears is a sign the roots are nearing exhaustion, and it is time to stop cutting.

Let the plumes grow all summer. In colder areas (zone 3 to 5), leave them standing to trap snow for better winter protection. In zones 6 to 9, cut the plumes off in September and cover the roots with 10 cm (4 inches) of coarse manure.

The first year of growing asparagus may seem like a fair amount of work, but once the beds are established, you will enjoy your own fresh asparagus for the next 15 years with very little care required.

Beans

I love my beans fresh off the vine, nice and crunchy. There are all kinds of bean varieties available today, but bush beans are one of the most trouble-free garden veggies you can grow. To get them off to a good start, beans need night-time temperatures of 10°C (50°F) or more, and they love sandy, loamy soils with no added lime. They mature in about 50 days from seed, and all you have to do is keep picking them to stimulate new flowers that will produce more beans.

Best green beans

- 'Blue Lake Bush'. Wonderful flavour. Ideal for a continuous supply. Prolific producer.
- 'Derby'. This award-winner has excellent flavour and very long, straight pods with little fibre.

Best yellow wax

- 'Nugget'. Long, round beans with very bright yellow colouring. A rich, buttery flavour. High yields.
- 'Valdor'. One of the most tender yellows. Very uniform, straight pods. Ideal fresh or preserved.

Best pole

Plant six beans per pole, and be creative with your designs—use a trellis or other framework. Although they're called pole varieties, they grow better up a string than on an actual pole. Keep picking regularly to keep the beans producing. They take about 70 days from germination before they're ready to eat.

- 'Blue Lake Pole'. More flavourful than their bush bean cousins, they produce very tender, meaty pods. Great any way you like them—fresh, canned or frozen.
- 'Kentucky Blue'. An award-winner, and not surprising with parents like 'Kentucky Wonder' and 'Blue Lake'. Long season production of very sweet, tender pods.
- 'Scarlet Runner'. Although not in the class of gourmet fresh beans, when cooked they are outstanding. I like to use them as ornamentals; their attractive red flowers look great on a trellis or chain-link fence.

Broad beans

Although not in the top ten choice of younger gardeners, fresh, tender broad beans smothered in pepper and butter are out of this world. They need staking, and you have to pinch out the centres when they mature to keep those tender leaves from attracting black aphids. Can be planted in November in zone 7.

- 'Broad Windsor'. The old standby. Long, full pods contain 6 to 7 easily shelled beans.

Beets

Beets are best as young plants and it's a real treat to thin them and enjoy the mild, sweet flavour of the tender leaves and small beet roots. Sow them early and every two weeks afterwards for a continuous crop until the real heat of summer. You can enjoy a last crop from a late summer planting. Good-quality roots are also great sliced, canned, pickled or frozen. They do best in a well-drained, but moisture-retaining soil with some manure added to the mix. They should never be allowed to be heat stressed. All can be harvested about 50 to 55 days after germination.

- 'Burpee's Golden Beet'. Golden roots are very tender and do not bleed like the red variety. The greens are wonderful in salads and taste like spinach when steamed.
- 'Formanova'. Has longer, carrotlike roots, providing more uniform slices than traditional round beets. Very high quality and ideal fresh, pickled or canned.
- 'Red Ace'. A hybrid of the well-known 'Detroit' types with very smooth exterior and deep red flesh. They have a delightful, honey-sweet flavour.

Brassicas (broccoli, brussels sprouts, cabbage, cauliflower, kale)

Early, midseason and late brassicas are available to extend the harvesting season of these valuable garden vegetables. Remember to plant them in the right sequence to prevent bolting.

There are two tips you shouldn't forget when growing all brassicas:

- Never grow brassicas in the same location as in previous years. Allow three-year intervals to prevent diseases.
- Keep them well watered and fed.

Broccoli

Broccoli is the most trouble-free of all brassicas, but it needs well-drained soil and a good watering at least once a week. As with most brassicas, there are early, midseason and late varieties adapted to spring, summer and fall weather conditions. The hybrid varieties are far superior to the older strains, but they all should be harvested continually to prevent flowering. On some varieties, cut the centre head out to encourage side shoots. Most are ready in 70 to 90 days.

- 'Paragon' (early). A Stokes seed introduction. Long, sweet spears with less fibre.

- 'Premium Crop' (mid). The main midseason variety, with larger flat heads that can be quartered.
- 'Packman' (late). Large, semidomed heads with lots of side shoots. Good for a wide variety of growing conditions. Sow in May or June.

Brussels sprouts

Always plant mid- to late June—sprouts taste better after a frost. Removing the leaves in late summer, when the sprouts are the size of peas, will make the sprouts larger, but the leaves do protect the sprouts from fall rains, so be careful. Harvest sprouts from the bottom up.
- 'Prince Marvel'. Small, wonderfully flavoured sprouts can be harvested October through December.

Cabbage

There are three distinct seasons for cabbage: early, medium and late. Plant them shallowly and don't cultivate too deeply. To prevent cabbage from splitting when the heads are mature, give them a gentle tug to break the hair roots.
- 'Emerald Acre' (early). Flavourful and good resistance to bolting. Compact, solid round heads.
- 'Gourmet' (mid). Round, blue-green, solid heads with short cores.
- 'Survivor' (late). Dark blue-green, dense heads with great disease resistance. They're good fresh and store well for a short time. They will tolerate fall rains and will sit in the garden up to 70 days without splitting.

Cauliflower

The goal is to grow pure white, crunchy heads that are ideal for dipping as well as creamed cauliflower. Early, midseason and late varieties extend the season. When heads begin to size up, fold or tie the leaves over the top to protect the heads from the sun and keep them white.

- 'Snow Crown' (early). Very uniform, 1-kg (2.2-pound), solid white heads. Mild, sweet flavour.
- 'White Sails' (mid). I love this variety because the leaves wrap themselves over the heads. Extra large, white heads.
- 'White Rock' (late). Well-wrapped leaves provide good rain protection for late harvests. Excellent late variety.

Carrots

For good root development these vegetables need a deep, sandy loam. They do not like manures. For healthy carrots, plant in a new location every year. Sow successive crops for a continuous supply into late fall. Flavour is my number-one priority with carrots, and that's what the following picks are based on. If you're after the best colour, plan to have your carrots mature when temperatures are in the 16° to 21°C (60° to 70°F) range.
- 'Baby Sweet Hybrid' (early). A high-quality mini-carrot with bright orange colour and delicious flavour.
- 'Earlibird Nantes' (early). One of the earliest hybrids with that incredible Nantes flavour. Bright orange colour and crunchy.
- 'Minicor' (early). Very sweet and crunchy with a rich orange colour. One of the best mini-carrots.
- 'Thumbelina' (early). A small round variety. Fabulous taste, great colour and can stay in the ground for a long time and retain its flavour.
- 'Napa' (mid). A Nantes type which pulls out of the ground easily. Deep orange colour, very crispy and good sweet flavour.
- 'Tamino' (late). Very sweet, cylindrical Nantes type with great colour. Keeps a long time in the ground, even through the winter in mild climates if you give them some protection.

Corn

I live in the self-proclaimed corn capital of Canada, and boy, do we know corn. There's so much information it's hard to keep up with all of the latest breeding programs. Your first decision is to choose between the different corn groups. You have "Sugary Normals" (su), the new "Sugar Enhanced" (se/SE) and the "Supersweets" (sh2). Each group has its own strengths and weaknesses, but it ultimately comes down to the flavour you prefer. Try not to mix these hybrids in your garden plot, as there is a risk of cross-pollination distorting the corn.

White kernels convert sugar into starch more slowly than the yellow kernels. This means white varieties taste sweeter for a longer period of time.

Good cultivation is as important as selecting the right variety. Plant when night-time temperatures are above 10°C (50°F). Corn likes to be grown in full sun in raised beds with good drainage. It likes soil with plenty of well-rotted manures. Plant after the last frost and in two-week intervals for a continuous supply in the fall. Corn is wind-pollinated and it's more efficient when plants are in blocks instead of long strips.

Name	Group	Days to harvest	Colour	Cob size	Comments
Ambrosia	SE	75	dark yellow	22 cm (8 ½ in.)	Very sweet, unique flavour.
Extra Early Super Sweet	sh2	67	yellow	28 cm (11 in.)	Superb flavour for an early corn.
Golden Jubilee	su	84	yellow	22 cm (8 ½ in.)	The real Chilliwack corn.
Milk 'n' Honey	sh2	71	white/yellow	22 cm (8 ½ in.)	Excellent colour contrast and wonderful flavour.
Phenomenal	sh2	85	white/yellow	22 cm (8 ½ in.)	Best late, and best-looking bicolour.
Sugar Finger	novelty	78	white/yellow	22 cm (8 ½ in.)	This novelty corn has such small cobs it's almost cobless, and you can eat the whole thing—honest.

Cucumbers

Cucumbers have been cultivated in India and China for about 3000 years, and Christopher Columbus is generally credited with taking them along on his voyages and introducing the plant to North America. However they got here, I'm a big fan of the cool, fresh taste of cucumbers in a sandwich, or preserved as a crispy pickle. They are a popular vegetable in the back garden and are easy to grow.

Cucumbers love well-rotted manure, so prepare your soil by digging in a healthy dose about 30 cm (1 foot) below the surface. This helps create a haven for the roots when the weather gets hot up above, providing moisture to keep the cukes from becoming bitter.

The most important thing to remember is that cold, wet ground is no place for any self-respecting cucumber. The most frequent mistake people make is to put them in the garden too early. Wait until late May or early June before setting out your transplants (which have been hardened off in a cold frame, of course). They need good drainage, so raised beds or bermed soil is important. Cucumbers are more delicate than many plants, so take extra care not to damage the roots when you're transplanting them.

Don't grow cukes on the ground where they are susceptible to all kinds of nasty things; instead make a wonderful display by taking a couple of trellises and securing them together as an A-frame. Grow the plants up one side and down the other for a terrific looking cucumber house. It keeps them off the damp ground and makes it tough for slugs to find them.

Wet spells in the summer or too frequent watering on the foliage can cause alternaria leaf spot, so water at root level rather than overhead. To avoid disease problems, move your cucumber patch to a new garden location every year.

There has been a lot of discussion about what makes cucumbers bitter, with no conclusive evidence pointing to a solution. It enters from the skin and travels to the centre of the cuke, particularly at the stem end. Slice into the ends of the cucumber until you're past the bitter part, cutting deeper at the stem end. For cukes that are least prone to bitterness, try burpless varieties, such as 'Sweet Slice', 'Sweet Success', and the long varieties 'Orient Express', 'Japanese Long Pickling' and 'English Telegraph'. If the plants are well cared for, well fed and mulched during extremely hot weather, and watered thoroughly and deeply, you've done everything you can to prevent bitterness.

Cucumber varieties

Pickling. The number of cucumber varieties is fast growing. 'National Pickling' and 'Straight Eight' remain popular, but I prefer some newer varieties. 'County Fair' is one of the first nonbitter pickling cukes. It's resistant to many diseases and produces heavy yields of seedless, white-spined pickling cucumbers. 'Double Yield Pickling' is the best variety for gherkins and dills, and has a rich green colour.

Slicing. For slicing cucumbers it's "burpless" or nothing; the burpless tag means you can eat the skin without fear of an upset stomach. The long English types are generally greenhouse grown, but you can have success outside with 'Japanese Burpless' or the shorter, but equally good American burpless, 'Sweet Slice'.

Containers. If you're growing in containers, try 'Pot Luck' or 'Patio Pic'. Be sure to use a minimum 35-cm (14-inch) container, full to the brim with a mix of good potting soil and sterilized compost.

Novelty. There are also 'Lemon' cucumbers, which grow the shape, size and colour of lemons, but taste like cucumbers. The New Zealand 'Crystal Apple' is another real novelty, with its small, rounded shape and prolific production.

Lettuce

While in Wisley Garden in England I was astonished at the lettuce displays. Move over annuals and perennials, and make way for lettuce! To add zip to both your garden and salad plate, grow some of the delicious and colourful red varieties.

Butterhead and Boston. 'Jacqueline' is an early Boston type, brilliant red, with some green showing. 'Red Vogue' is a smaller Boston type with heavy, rosy red shades on a light background. 'Sangria' is a thick-leafed, medium-sized bicolour butterhead. Leaves are tinted with a warm, rosy red on a mid-green background. 'Bicolor Novelty' has fine-cut leaves tipped with dark red.

Looseleaf. 'New Red Fire' is bright red in summer, and turns to dark red in fall. 'Red Sails' is deep bronze-red with savoyed (puckered), open leaves. 'Ruby' is the darkest red, with savoyed leaves. 'Impuls' is the darkest burgundy, with excellent flavour.

Mesclun. If you want it all in one packet, grow your own salad with one of the Mesclun mixes that matures in just 20 days. It includes 'Salad Bowl' and 'Green Rapids' lettuce, cos, 'Pak Choi', 'Hon Tsai Tai', kale and brassica greens. All you need is the olive oil and vinegar and you're set.

Peas

Does any vegetable taste better than fresh peas right out of the pod? This is something you simply can't buy. Peas are cool-weather plants and need to be sown early to mature in June and early July. Second fall crops can often be sown in early August. Peas detest wet feet. The enation virus, which is spread by green aphids, often attacks and destroys pea crops when the warm weather hits in early summer, so keep your peas free of aphids.

Edible pod peas. Even when these peas are old, the flavour remains good. Ideally, they should be harvested when you can see the outline of the young peas.

- 'Sugar Snap'. Needs a trellis or support because it grows up to 2 m (6 feet). Pods are sweet and crunchy.
- 'Sugar Ann'. An award-winner and the compact version of 'Sugar Snap'. Vines that grow to 60 cm (2 feet) produce smaller pods but are just as flavourful, and they are more versatile in the garden.

Shelling peas.
- 'Alderman' ('Tall Telephone'). An oldie, but still one of the best-tasting tall peas. Large pods produce lots of delicious peas for continuous harvesting.
- 'Maestro'. Very high-yielding, compact 60-cm (2-foot) plants with outstanding flavour.

Peppers

These are the hot vegetable in recent years, in a number of ways. The hot and spicy ones really jazz up your food. The sweet ones add colour and flavour to salads and hot foods alike. They are marvelous for stuffing, pickling, making relish and salsa or simply eating fresh off the vine. The incredible colour range—white, purple, orange and even chocolate brown—adds life to ordinary-looking meals.

Hot peppers. Some people swear by scorching hot peppers. They say these mouth-numbing zingers help you sweat the toxins out of your body. An interesting fact about peppers is that their heat is multiplied by a factor of 10 once they are dried, so if you want to boost those sweat-inducing qualities, put your peppers on the drying racks. There are specialty publications devoted to peppers, and there seems to be a subculture devoted to seeing who

---

*T*ips for growing great peppers

- Set out transplants only when the weather is warm. These are definitely not cold-weather plants! Night temperatures should average 17°C (62°F).
- Raised beds and well-drained soil are vital to well-developed root systems—the key to pepper success.
- Side-dress with calcium nitrate to promote that all-important plant root growth.

can handle the hottest of the hot ones. This list ends with the hottest of the hots.

- 'Pimento Elite'. Thick-walled fruits turn bright red when mature. Mildly hot; great for salads.
- 'Señorita'. Smooth green bell-type pepper turns red when ripe. Great for salsa, stuffed peppers, or fresh in salads.
- 'Hungarian Wax'. Bright yellow to red. Very productive; used in European cooking.
- 'Garden Salsa'. Dark green to red. Ideal for salsa.
- 'Long Thin Cayenne'. Green to red, 15 cm (6 inches) long by 130 mm (½ inch). Very good for drying.
- 'Early Jalapeño'. The best known of the hot ones. Green to red, good fresh or pickled.
- 'Super Chili'. Green to red. Spicy, cone-shaped chili peppers.
- 'Super Cayenne'. 10 cm (4 inches) by 2.5 cm (1 inch).
- 'Tabasco'. Green to yellow. Very pungent.
- 'Habañero'. No fooling around here. If you touch even the foliage and then your face, it will cause a burning sensation. 2.5 cm (1 inch) by 4 cm (1½ inches), green, thin-walled fruit that wrinkle at maturity. Don't say I didn't warn you about this one.

Sweet peppers. Many green varieties will turn red if left on the plant to mature. The longer they mature, the sweeter they get. Here are a few personal favourites.

- 'Banana Supreme'. About 20 cm (8 inches) long by 5 cm (2 inches) wide. Sweet, yellow banana-type. Big plants produce large fruits.
- 'Better Belle'. Best of the large, green bell-type peppers.
- 'Gypsy'. Yellow bell, sweetest of all peppers. Up to 14 cm (5½ inches) long and 8 cm (3 inches) wide.
- 'Jingle Bells'. Perfect for Christmas. Small, 4-cm (1½-inch) square fruits are versatile and ideal for stuffing and broiling.
- 'Sun Belle'. Very compact plant produces large, 10- by 13-cm (4- by 5-inch) fruits.

Potatoes

For a favourite fresh vegetable, I'm torn between new peas and tiny succulent potatoes smothered in butter, so I enjoy them together.

There are early, midseason and late varieties of potato, but every year the race is on for the first new potatoes. Small seed potatoes are your best buy. To have success with spuds, grow them in raised beds and plant them shallow to keep them warmer and drier, especially early in the season. The roots will quickly spread into the bed and begin producing young potatoes.

Remember, to keep your potatoes healthy and free from scab, do not add lime, fresh manure or wood ash to the soil.

Best early potatoes (60 days)

- 'Caribe'. A blue-skinned, scab-resistant variety with white flesh. Outstanding uniformity. Great for boiling.
- 'Norland'. A smooth-skinned red with shallow eyes that is good for boiling and French fries.
- 'Warba'. A golden-skinned, flavourful spud with deep eyes that is used for boiling and baking.

Best midseason potatoes (75 days)

- 'Desiree'. This unique combination of red skin and yellow flesh is one of the hot newer varieties from Holland. Very high yielding and good flavour.
- 'Kennebec'. One of the higher yielding, very large varieties with smooth yellow skin and white flesh. Great flavour and stays firm when boiled. Also good for baking and French fries.
- 'Pontiac'. Perhaps one of the most popular potatoes. It's a mid to late potato with red

skin, white flesh and great flavour. Most often used for boiling.
- 'Yukon Gold'. One of the best baking potatoes with attractive yellow skin and flesh. Shallow eyes. Also good for French fries.

Best late potatoes
- 'Russet Burbank'. High-yielding, long potatoes with light brown skin and white flesh. Great for baking and stores well.

Pumpkins

Let me put in a plug for this North American native, the lowly pumpkin. We tend to look at them only for jack-o'-lanterns, but pumpkins have a lot more to offer. Those with young families should have a few in the garden because they're perfect for introducing a child to horticulture. In the Middle East the seeds are thought to enhance energy.

Seeds from varieties such as 'Hull' are wonderful baked or roasted, and seeds can be dried and put in the freezer over the winter. The pumpkin with the best-tasting seeds is 'Lady Godiva'. Of course the meat is good for baking pies and cookies and as an ingredient in soups. For pies, I recommend 'Sugar Pie', which grows up to 6 kg (13 pounds), or 'Lumina'.

Pumpkins used to take up a lot of space, but newer varieties such as 'Spirit' are suitable for smaller gardens. 'Sweetie Pie' is a miniature that fits in the palm of your hand, and weighs in at 150 grams (5 ounces). Miniatures can be eaten, or dried and waxed for use as a table decoration.

Of course, many people like to grow pumpkins for Halloween carving. Here's how to grow a really big one.

1. Start by choosing a big variety, such as 'Prizewinner' or 'Big Max', at 45 kg (100 pounds). These still look like pumpkins, unlike the ugly world record holder for size, 'Dill's Atlantic Giant', which weighs up to 400 kg (900 pounds) and looks like a yellow-orange squash with no particular shape.
2. Prepare the soil with well-rotted manure, dug in deep. Plant the seeds in a raised bed.
3. Remove all fruits but one or two on each plant. This will put the growing efforts into these remaining pumpkins.
4. Do two plantings, about one week apart.
5. Keep plants mulched in hot weather; extreme heat can lead to a 25 percent reduction in size.
6. Pinch back extra vines and continue removing new fruit.
7. Feed with 10-14-21 fertilizer.
8. Harvest with at least 8 cm (3 inches) of stem still attached to the pumpkin.
9. Bring in the front-end loader to lift the beast out of the garden.

Squashy gourds and gumpkins

"Squashy gourds" and "gumpkins" are the result of growing ornamental gourds, squash and pumpkins in the same vicinity. To prevent ornamental gourds from crossing with late squash and pumpkins, separate them by at least 240 m (800 feet). It's a long way, but it's necessary. Early squash and gourds don't seem to cross-breed.

To dry gourds and mini-pumpkins for ornamental purposes, pick the fruit carefully so that the stems are intact. Wash with a strong disinfectant to remove soil and fungi. Spread them out to dry in a well-ventilated area for about a month. When they are dry, wax them with regular floor wax and give them a polish with a cloth.

If you're not going for enormous size, and want a good Halloween pumpkin, try 'Connecticut Field', which grows up to 11 kg (25 pounds) or 'Howden', which is slightly larger.

There are a few white, extra-spooky types, such as 'Casper', which grows to about 5 kg (11 pounds) or 'Lumina', which is slightly larger. For smaller pumpkins, try 'Little Lantern' or 'Spooktacular'.

Spanish onions

These are the huge, sweet, tear-inducing guys that taste so wonderful in sandwiches or fried. Plant them by March 1 at the latest to make sure they are sweet by harvest time. They must be cured in the sun at least a week for good storage. Most mature in September.

We seed them into starter trays in late January, harden them off in a cold frame in late February, and as soon as the real hard frosts are over, into the ground they go. Cut the tips of the young plants back to get the root system well established and use the mud slurry described in the transplanting seedlings section of chapter 1.

Grow in raised beds with sandy, well-drained soil to minimize root maggot problems. Rotating crops to different areas and using a row cover, such as Reemay cloth, will make a significant difference. Usually the adult flies lay their eggs in early to late summer near the base of the onions. The cloth prevents this from happening. Brown pupae overwinter in the soil, but fall rototilling will help bring them to the surface where they will die.

These are the best Spanish varieties I've found, from giant to huge.
- 'Gringo'. Beautiful copper colour, sweet flavour; 105 days to maturity, 13 cm (5 inches) across.
- 'Kel Sae Sweet Giant'. World Record Onion at 3.368 kg (7 pounds 7 ounces). Mild sweet flavour; 110 days to maturity. Grows easily in almost any climate and will keep until early January.
- 'Riverside Sweet Spanish'. The workhorse of Spanish onions, sweet and mild; 115 days to maturity, 13 cm (5 inches) across. Will store until January.
- 'Walla Walla'. One of the sweetest of all Spanish onions; 95 days to maturity, 13 cm (5 inches) across. Only keeps until late December, but chances are you'll have eaten them all by then anyway.
- 'White Sweet Spanish'. The nicest white Spanish; 120 days to maturity, 14 cm (5 $\frac{1}{2}$ inches) across. Stores well.

Summer squash

If you're not already growing these vegetables, you don't know what you're missing. Some varieties mature in just 55 days, and they are wonderful in stirfries or salads. Marrow types are bland on their own but are great stuffed with ground meat, hot peppers and herbs.
- 'Scallopini' is good for slicing, dicing and stir-frying.
- 'Vegetable Spaghetti' takes about 70 days to mature. Just cut it in half, remove the seeds, and steam, bake or microwave until it's tender. The flesh separates into spaghetti-like strands—nutritious, delicious spaghetti with low calories to boot.
- 'Zucchini Select' is a bush zucchini that limits its production of fruit to under a million (more or less). Harvest it while the flower is still on the fruit (avoid those four-footers).

Tomatoes

This plant has come a long way in the past hundred years. It's only in this century that tomatoes have become a staple food in kitchens around the world. They were widely cultivated by early inhabitants of the Americas before being brought to Spain and dubbed *tomates* based on the word in the Aztec language. They belong to the nightshade family and, as such, were negatively associated with their poisonous relatives, belladonna and deadly nightshade. But tomatoes have shed that reputation and have been embraced in a great many forms, from fresh produce to ketchup.

Of all the edible plants in the garden, tomatoes are the most popular. Over the years I have received a lot of frantic calls from people who are having problems and want to save their plants. There are a number of common problems that can be avoided or solved quite simply.

Start by selecting healthy-looking plants with good strong stems and thick foliage. And don't be in too big a hurry to get your tomatoes in the ground as soon as it warms up in the spring. Hold your horses! Transplants shouldn't go out until late May or early June in most areas. If the leaves turn purple, that means the plants are too cold. The cold will not permanently harm them, but it will stunt the growth. If you want to get a head start, plant them in 4- or 8-litre (1- or 2-gallon) containers and grow them in a cold frame until it really warms up in late May. Even simpler, you can use inexpensive, wire tomato cages wrapped in polyethylene to create instant mini-greenhouses. (Another reminder you don't have to be wealthy to garden, just creative.)

Plant tomatoes in the warmest part of your garden. They will do especially well up against a south-facing wall or fence. Tomatoes send out large roots that draw a lot of moisture and nutrients from the soil. Work the soil at least 45 cm (18 inches) deep, and mix in plenty of well-rotted manure. Be sure to add a handful of dolomite lime, which provides calcium to prevent blossom end rot. Some people place shredded newspapers in the planting holes to hold extra moisture during those hot summer days.

Plant tomatoes in raised beds or with soil bermed up around them to give extra warmth. Commercial growers often place black plastic over the berms to concentrate the heat. Initially, feed them with fine bone meal. Superphosphate 0-18-0 will get the roots off to a fast start. A root stimulator will also give them a quick boost. A burst of foliage in the first few weeks will get the plant growing, but later in the season you want the energy to go into fruit production. To this end, start with a well-balanced fertilizer, such as 20-20-20. Once the plant begins to flower, use a low-nitrogen fertilizer, such as 4-10-10 or 0-10-10. This will help create a compact plant with more flowers.

Cage them or stake them but whatever you do, give your tomato plants something to hold them up. Pantyhose is great for tying tomatoes to stakes. It supports them without cutting into the stem of the plant. Alternatively, you can purchase inexpensive vinyl ties. Just be careful what you use to tie up your plants;

Container tomatoes

If you're growing tomatoes in containers, use sterilized potting soil for outside containers. A minimum size pot of 26 litres (7 gallons) will allow the roots to grow large enough to support the plant. Every time I see a tomato plant in a 4-litre (1-gallon) container I renew my vow to organize a Tomato Liberation Front and set those roots free.

The most reliable and popular varieties for containers are cherry tomatoes, such as 'Sweet 100' or 'Sweet Million'. They are indeterminate, which means they keep growing and producing all through the summer. Many of the smaller, early varieties are also good in containers.

Tomato problems

- Early blight and late blight are caused by two different fungi and create common and serious problems for tomato growers. They usually show up after cool, wet weather and can rapidly wipe out your crop. The first sign of early blight is small brown dead spots on the older, lower leaves. Older fruit develops dark, leathery, sunken areas. Late blight causes small, dark, moist spots on leaf veins and/or stems. Ripe fruit develops light-coloured patches. If you see signs of blight, remove the affected areas immediately. Do not dispose of infected plants in the compost—this is a water- and air-borne fungus and could infect next year's crop. Do not plant tomatoes in the same location two years in a row.

  Copper spray helps prevent blight. It works better if you mix it with warm water rather than cold. For information on spraying with copper, see the licensed herbicide dispenser at your local garden centre.

  Once blight enters the stem of tomato plants, it is pretty well game over. The only thing to do at that point is pick the green tomatoes and allow them to ripen indoors. Dip them in water mixed with a capful of bleach to kill fungus and bacteria. Rinse them and allow them to ripen in an area with good air circulation.

  The best way to deal with blight is to prevent it in the first place, by growing under protective covering. I recommend plastic bags specially designed to go over tomato plants. They are tinted so that the sun keeps them warm but will not burn them through the protective plastic. They will ripen faster as well. Make sure you pull open the bags at the bottom so pollinating insects can still get in.

  If you insist on leaving your plants uncovered, then keep the foliage as dry as possible—no overhead watering!

- Droopy plants are an indication of stress, often because of hot weather. To prevent heat stress, mulch them to keep moisture in the soil. Water them in the morning and water thoroughly. You want to encourage those roots to grow down.

- Tough tomato skins may be caused by a lack of micronutrients. Mix fritted trace elements with water, give the tomatoes a good drink and then lime the area. You will only need to do this once. As they are growing, give them magnesium in the form of Epsom salts. Use all of these and you will cover all of the trace elements. In addition to these goodies, keep feeding with the usual tomato food.

- If the bottoms of your tomatoes turn black, you have a calcium deficiency known as blossom end rot. To prevent it, simply add a few handfuls of dolomite lime to the soil at planting time. Once the fruit has blossom end rot, you can't do much about it in the current growing season.

don't use anything like string or twist ties, which will cut into the stem. Many tomato varieties grow surprisingly big if they are well supported, so use extra-long stakes—at least 2 m (6 feet). Those small wire cages should be banned as tomato supports—they're simply not large enough. You can make your own industrial-strength version by buying concrete reinforcing wire and fashioning it into round cages hefty enough to support 12 kilograms (26 pounds) of fruit with no problem.

To ensure your flowers turn into luscious fruit make sure the plants are pollinated. Attract pollinating insects (see "Companion planting" later in this chapter) or simply go out and gently shake the vines.

If growing tomatoes in a greenhouse, make sure you keep it open so pollinating insects can get in to do their thing. When it gets very hot and dry, you can hang a bit of Reemay or shade cloth to cut the intensity of the sun. This will also help avoid those dreaded spider mites. 'Moneymaker', 'Vendor' and 'Early Girl' are good hothouse varieties.

You can save tomato seeds of nonhybrid varieties for planting next year. Let the tomato mature until it is overripe, almost to the point of rotting. Then break it open and take the seeds out. Put them onto a tissue and let them dry. When the pulp is completely dry, collect them in an envelope and put them in the deep freeze or refrigerator until next year.

Tomato varieties

Everyone who grows tomatoes has a favourite variety, but I really encourage people to make room for some of the new types. It never hurts to compare, and you might be lured away from some of those old favourites once you try the new ones. Although they haven't yet developed a blight-free tomato, breeders have succeeded in overcoming many of the diseases that plagued older types. You will often come across tomatoes tagged with some or all of these letters: V, F, N and T. This means disease resistance to, respectively, verticillium, fusarium, nematodes and tobacco mosaic.

There are many varieties out there, but often garden stores only stock five or ten of the hundreds available. Shop around for some of the varieties below, or start your own tomatoes from seed.

Tomato type	Name	Days to harvest	Comments
Early	'Early Girl'	52	113- to 170-g (4- to 6-oz.) fruit throughout the summer.
	'Early Girl Improved'	52	Even more vigorous, with much more fruit.
	'First Lady'	55	Similar to 'Early Girl' but even more productive and flavourful.
	'Beefeater' or 'Italian Beefsteak'	60	Novelty varieties producing massive, 900-g (2-lb.) fruit. Outstanding flavour.
	'Ultra Girl'	62	Nice 225-g (8-oz.) fruit.
	'Champion'	62	Round and blocky fruits, weighing about 285 g (10 oz.). Produces all summer long.

Tomato type	Name	Days to harvest	Comments
	'Ultra Sweet'	62	Wonderful flavour, with a nice balance of acid and sugar. Grow to about 285 g (10 oz.).
Midseason	'Ultrasonic'	65	Tremendous quantities of 170-g (6-oz.) and larger fruits, under a wide variety of conditions.
	'Super Fantastic'	70	Slightly larger, 285-g (10-oz.) fruit.
	'Celebrity'	70	Very robust plants with 285-g (10-oz.), dark red fruit.
	'Floramerica'	75	Lots of very tasty fruit in the 255-g (9-oz.) range.
	'Oregon Spring'	75-80	Not the biggest crops, but one of the biggest flavours. Pretty hefty size, quite early.
	'Roma'	75	Probably the most productive for its size. Tremendous amounts of 85-g (3-oz.), pearlike fruits.
Large	'Beefmaster'	60	Very large, 500-g (18-oz.) delicious fruits that are resistant to cracking and splitting.
	'Pik Red'	60	Best of the west coast beefsteak types, with tasty, huge fruits.
	'Ultra Magnum'	68	Short, stocky plants produce 400-g (14-oz.), smooth fruit with a sweet flavour.
	'Better Boy'	72	Smooth, round, flavourful fruits up to 450 g (1 lb).
	'Ultra Boy'	72	Perhaps the best of the 'Boy' Series, with 450-g (1-lb.) fruits that are deep-globed and produce all through the summer.
Cherry	'Tiny Tim'	45	An oldie but still a great compact plant with loads of cherry-sized fruit.
	'Sweet 100'	60	A proven performer, with loads of sweet fruit on fast-growing vines.
	'Sweet Million'	65-75	Even better. Fast growing, sweetest of all cherry tomatoes and highest in vitamin C.
	'Sweet Chelsea'	67	The jumbo of the cherry tomatoes. 28-g (1-oz.) fruits on 1-m (3-ft.) plants; heavy production.
Patio	'Red Robin'	56	One of the best for containers.
	'Lunch Box'	62	Very flavourful, high quality, egg-sized fruit. Extremely productive.

Tomato type	Name	Days to harvest	Comments
Low-acid	'Sunset Gold'	60	A heavy-producing yellow cherry that's sweet and low in acid.
	'Orange Queen'	65	170-g (6-oz.) beefsteak-like fruit.
	'Lemon Boy'	72	A distinct lemon yellow with wonderful flavour from the 200-g (7-oz.) fruit.
	'Yellow Stuffer'	80	Looks like a bell pepper. Attractive, 170-g (6-oz.) fruit.
Novelty	'Tumbler'	50	Very sweet, cherry-sized fruits will droop and hang from baskets. Side shoots also produce.
	'Longkeeper'	70	Mid-sized orange-red fruit. If picked half ripe before the first hard frost, it will slowly ripen over 12 weeks in a cool, dark place.

Organic Insect and Pest Control

Organic insect control doesn't have to mean extra work; much of it is simply commonsense gardening.

- Keep a weed-free garden. Weeds are a prime habitat for insects during different stages of their life cycles. If you don't like weeding, use the effective and inexpensive weed barrier cloth. Once you try it, you'll wonder how you gardened without it.
- Make sure your soil is in top shape. I'm always going on about light and sandy soils, and with good reason. They are far less inviting to many insects, which prefer heavy, wet soils. Sand and bark mulch or sawdust (fir or hemlock only) go a long way to lightening heavy soils.
- Be sensible about garden planning. If you are having trouble with cabbages, radishes and turnips, why grow them? Have the vegetables you enjoy, but concentrate on those with fewer problems. Pole beans, broad beans, parsnips and cucumbers are usually resistant to insects. (Be more selective in the flowers you plant as well. When is the last time you saw insects or slugs attack fibrous begonias or impatiens?) If you have a disease problem in one area of your garden, don't plant the same crop in the same location next year. Many diseases stay in the soil and will be picked up again the following year.
- Follow the companion planting guidelines at the end of this chapter. If you're not sure, just plant garlic with everything!
- Use row covers, one of the best weapons in the bug war. They are cheap, easy and effective. Drape Reemay covering over young seedlings, leaving plenty of room for expansion, hold it in place with soil, and it can remain until the plants mature. The plants will be warmer and grow faster and further into the season, and insects won't be able to get at them. Cutworms, for example, result from eggs laid on plants by a moth. The only downside to protective cloths is that they might cut down on air circulation and increase humidity, leading to disease problems.

- Organic pesticides may be part of the solution. B.T. (*Bacillus thurengiensis*) is one of the best controls for chewing insects. It has to be used about once a week when those moths and white butterflies are flitting about, but it works. If you choose to use B.T., or any other insecticide, read the directions carefully.
- Many places sell ladybugs by the package, and they're happy to eat those annoying aphids. Another weapon in the arsenal of living things used to control nasty pests is friendly nematodes, a type of parasitic worm. We've had pretty good success with them in the control of slugs, maggots and insects that spend part of their life cycle in the soil. They will last up to 15 months in the soil as long as it stays moist and temperatures don't dip below −18°C (0°F). Release them at night, allowing them to get into the soil without sunlight hitting them. Water them into the soil and let them get to work.
- Plant flowers and shrubs that attract birds, and install a birdbath or two. This may be your best form of insect control.
- Silicon dioxide, otherwise known as diatomaceous earth, is a powder made from deposits of ancient creatures called diatoms. Many companies are producing it under different labels for use in controlling crawling insects.
- There are many soap products available, and the broad-spectrum types, often including pyrethrins, seem to be the most successful. Pyrethrins and many other products have been available for years, and they do a pretty good job.

The secret for any of these methods is careful observation. When your plants are half-eaten and the insect population is booming, it's already too late. The time to catch them is when they're on the rise; hit them before they get out of control. Make a note on your calendar if you have problems, and next year, begin taking preventive measures a couple of weeks ahead of time.

Companion planting

Companion planting is a great idea that's being rediscovered and improved upon in many quarters. The idea is to group together plants that aid each other's growth and help protect from disease and insects as well. There is not a lot of scientific research in this area, but in my experience it results in better growth and fewer insect problems.

One of the most important reasons to use this method is to attract pollinating insects to the garden. Tomato flowers must be pollinated continuously, even in rainy weather, or we simply won't get any fruit. Fragrant heliotropes or blue petunias planted nearby work like magic to attract bees, even on cloudy days.

Beneficial insects can also be attracted to ward off problem insects. Ladybugs won't be too far behind a heavy infestation of aphids, but to attract them earlier a few special weeds, like lamb's quarters (*Chenopodium album*), certainly help. This plant is widely known as an annual weed that can spread quickly, but in small numbers it's a benefit to any garden. A close relative of spinach, it tastes just as good and is even richer in vitamin C.

Many plants can act as insect decoys to protect other plants. Nasturtiums, for instance, attract aphids, so well that apple growers plant them under their trees to keep woolly aphids off the trees. If sown in a greenhouse they help repel whitefly. If you're growing nasturtiums just for their gorgeous flowers and aphids are a problem, simply dust the nasturtiums with lime to discourage the aphids.

You can't talk about companion planting without mentioning garlic—one of the best repellents you'll find. Grown near roses, it is said to repel aphids. Around tomatoes, it may help keep spider mites away during dry, hot summers. You do have to use some caution with garlic, though. It can inhibit the growth of both peas and beans.

Parsley is supposedly good for repelling carrot rust fly. Plant it in the same row as carrots. Parsley also seems to give extra vigour to asparagus and tomatoes.

Certain plants develop a synergy when grown together. Basil, for example, enhances plants growing near it, with the exception of rue. Radishes and carrots grow well together, as do corn and lettuce.

Some plants are allelopathic, which means they adversely affect the growth of nearby plants. For example, sunflowers and walnut trees have properties that are phytotoxic to other plants. Some plants are bad together because they are susceptible to the same insect or disease. For instance, potatoes and tomatoes should never be planted near each other because they are both susceptible to blight; if one crop gets blight, the other won't be too far behind.

Herbs and Edible Flowers

When it comes to herbs, many people can't get past the question of whether or not to pronounce the "h." Is it "herb" or "erb"? No matter how we say the word, herbs have exploded in popularity in recent years. Basil used to be found growing almost exclusively in close proximity to a person of Italian heritage, but now it's shooting up in containers and gardens all across the land. Herbs combine two of the strongest interests in North America at the moment—gardening and cooking, both of which are becoming more sophisticated. Where would our tables be without basil pesto, thyme and garlic herb butters, cilantro salsas or the traditional sage-scented stuffing in the holiday turkey?

We are seeing a resurgence of interest in natural remedies using herbs and herbs also play a role in conventional medicine—hundreds of modern drugs contain plant extracts as active ingredients and more are being tested and marketed every year.

In our gardens, herbs, like many vegetables, cross boundaries between the practical and the aesthetically pleasing. We grow them not only to harvest, but also for their attractive foliage, flowers and fragrance. Herbs cover a wide range of plants, and some of the plants we call herbs are also mentioned on other pages of this book. Richter's herb catalogue, which grows fatter every year, contains many interesting new herbs as well as old favourites.

Dandelions, nasturtiums and lily-of-the-valley are all classified as herbs. The American Herb Society defines a herb as "any plant that can be used for pleasure, fragrance or physic." That certainly opens the door, doesn't it?

Don't be intimidated by the list, though. Start by growing the herbs you enjoy using for culinary purposes, or for their flowers and fragrance. Based on my own experience, you'll soon begin to experiment with more and more herbs.

Herbal Basics

Most herbs need a site with well-drained soil, where they get at least five to six hours of sunshine per day. Moderate garden soil enhanced with organic matter is fine, but heavy applications of fertilizer, especially those with high nitrogen, will decrease the essential oils in the plants. Keep the pH in the range of 6.5 to 7.5, which may require the addition of dolomite lime in winter. The good news is that herbs usually require only a little bit of maintenance once established.

Herbs can be classified as annuals, biennials, tender perennials and perennials. It's important to understand this in planning your herb garden so you know which plants will survive the winter, which ones should be brought inside for winter protection and which ones you have to say goodbye to each fall. Herbs are started from seed, layering, cuttings or root divisions. Many are easy to start from seed and some, such as borage, anise, chervil, coriander, dill, fennel and parsley, should be sown only where they are to grow because they do not transplant well.

Harvesting and preserving herbs

Did you know that herbs grow better if they are harvested frequently? Start by pinching back young plants and continue to bulk them up with pinching and harvesting. The best time to harvest is in the morning when the dew has dried but before the sun gets too hot. Most herbs should be harvested before flowering to maintain production. Frequent cutting may delay flowering, but flavour and oil concentration are at their peak when flower buds first appear. The leaves can be used fresh or preserved in many ways for later use.

Annual herbs can be harvested right up until the first hard frost, while most perennials are best clipped from late spring until flowering. Stop pruning perennial herbs about one month before the first heavy frost so you don't encourage late soft growth.

Herbs in containers

Herbs are natural candidates for growing in containers. My longtime favourite has been a combination of chives and parsley. Chives create a nice focal point in the centre of a 25- by 30-cm (10- by 12-inch) hanging basket, and the curly leaves of parsley look especially attractive drooping over the edges. Both can be harvested all summer and through the winter in mild regions.

If it's colour you're looking for, do a potpourri basket. Add some purpleleaf 'Opal' basil and golden marjoram to the chives and parsley. Rosemary has attractive silver foliage and fragrant leaves, as does santolina. Some of the variegated mints and thymes also add that special touch for colour, fragrance and table use.

Freezing

Believe it or not, freezing is the easiest way to preserve herbs. Wash the herbs quickly in cold water, give them a shake, then chop them with a very sharp knife to preserve the oils. The clever way to freeze them is to drop a big pinch in each compartment of an ice cube tray, fill with water and freeze. When the cubes are frozen solid, place them in a freezer bag. To use them in cooking, drop the ice cube in the dish you're preparing and *voila!* The herbs will taste and smell as good as the day you picked them, although they won't look quite as good.

Drying

Drying is the more traditional way of preserving herbs. Pick them in the morning, tie them in small bunches and hang them upside down in a warm, dark room with very good air circulation. For leafy herbs, strip the leaves and spread them out on screens in the same conditions. Turn them frequently. Herbs used for crafts are best dried in silica gel or kitty litter, which takes out the moisture very efficiently. But don't use herbs dried this way for cooking.

Store the herbs in airtight containers, and keep an eye on them. If you notice any sweating or rotting, repeat the drying process. Herbs stored this way will last for about a year before they lose their flavour. Dried herbs have a far more concentrated flavour than fresh or frozen, so if you are using fresh herbs, you may have to use double or triple the amount.

Culinary Herbs

Every person's list of herbs is tailored to personal taste. Following are some favourites.

Annuals and tender perennials

The following herbs need to be planted every year, as they will not overwinter.

Anise (*Pimpinella anisum*). Annual. Grows to 60 cm (2 feet). Sow seed in spring when the ground is warm. It needs full sun and well-drained soil. It blooms in summer and flowers quickly, so keep an eye on it. Save the seed heads and dry them upside-down in paper bags; the bag will catch the seeds as they fall. Steeped in boiling water, the seeds make a refreshing anise tea. They also add a distinctive taste to baked goods.

Basil (*Ocumum basilicum*). Annual. Basil should never be planted outside until June, at the earliest. It needs heat, well-drained soil and hates being overwatered. It grows to 60 cm (2 feet) and blooms in July and August. Pinch it back often to keep it bushy. Protect it from slugs, which love basil. Grows well in big containers on a deck with lots of light. Watch out for *Fusarium* wilt, a fungal disease that produces black patches on leaves and stems. There is no cure; remove and destroy the plants.

- 'African Blue' basil. This unusual basil is sweet and has a bit of a camphor scent. It has attractive purple leaves and flowers almost constantly, with purple flower spikes. It is a tender perennial, and is the one basil that, with lots of light, does well indoors in winter.
- Cinnamon basil. It always looks great with its continuous-blooming, long, purple flower spikes. It has a cinnamon taste and perfume.
- Lemon basil. Its intense lemon perfume is ideal for teas. 'Dani' is the 1998 award winner.
- 'Siam Queen'. This award-winning basil was herb of the year in 1997. Its fabulous burgundy-purple flowers match deep green leaves with a purple hue. It has a spicy, licorice aroma and taste.
- 'Special Select' Genovese basil. This variety from Richter's does not have the mint overtones of other varieties and is one of the best for pesto.

Bay laurel (*Laurus nobilis*). Tender perennial. Generally hardy in zone 8, it will only take about 10 degrees of frost. It can grow to 12 m (40 feet), but reaches 1.5 m (5 feet) in a pot. It needs well-drained soil and sun to partial shade. Buy plants or start from cuttings. Keep an eye out for scale and control with horticultural oil. The tree looks great, even as a small plant, and the fresh leaves add flavour to soups and stews.

Cilantro, leaf coriander or Chinese parsley (*Coriandrum sativum*). Hardy annual.

Fragrant oils

The flavour and fragrance of herbs comes from the oil in the leaves. You can smell this evaporating oil in the air when the leaves are crushed, or even brushed with your hand.

Seeds work better than transplants, and can be sown when the soil temperature is 13°C (55°F). It loves light, well-drained soil. Grows .6 to 1 m (2 to 3 feet) tall and blooms in late summer. The leaves complement poultry, salads, salsas and soups. Save the seeds to flavour stews and pastries.

Dill (*Anethum graveolens*). Annual. Where would pickling cucumbers be without dill? This airy plant grows to 1 m (3 feet) and needs full sun and well-drained soil. Direct-seed it (you'll find it self-seeds very easily). Blooms midsummer. I love its perfume and use it as a cut flower mixed in bouquets with other summer flowers. Plant the dwarf fernleaf variety in pots for a little spice on your patio.

Lemon verbena (*Aloysia triphylla*). Tender deciduous shrub. Hardy to zone 9, it may tolerate temperatures as low as –9°C (16°F). Start with plants or take cuttings. It needs rich, moist soil and full sun. Grows to 1.5 m (5 feet). Watch out for whiteflies and spider mites. This South American native has an effusive lemony fragrance that puts lemons to shame. Its strong flavour makes a good tea, or you can use it as a garnish in iced drinks. The leaves are wonderful in potpourri, and a sachet in with your socks and linens will keep them smelling fresh.

eed collection

Collecting seeds from herbs is the same process as with other plants. Wait until they are mature; they should be completely dry and ready to fall. Dry them in a room with good air circulation or out in the sun. When they are thoroughly dried, place them in sealed plastic bags, label and pop them in the deep freeze until you need them.

Rosemary (*Rosmarinus officinalis*). Tender evergreen shrub. Hardy to zone 8, it grows up to 1.5 m (5 feet). Buy plants or start it from cuttings. Rosemary needs a very well-drained location and once established will tolerate drought. Blooms in early winter. 'Arp' is the hardiest variety, tolerating –20°C (–4°F), but my experience is that –11°C to –12°C (10° to 12°F) is the limit even in well-drained, protected areas. Most other varieties are less hardy and should be brought inside for winter protection. Pick it fresh for use in flavouring meats, particularly stews, and vegetables. It's a nice garnish and great for potpourri. Do not remove more than 20 percent of the plant at one time.

- Best trailing varieties are 'Santa Barbara' and 'Huntington Carpet'. They are very prostrate and full, earlier to flower and faster growing. 'Huntington Carpet' has a great trailing form ideal for hanging baskets.
- Best upright varieties are 'Arp' (the most winter hardy, to zone 6 with some protection), 'Rex' and 'Tuscan Blue', which have large dark leaves and a vigorous growth habit.

Summer savoury (*Satureja hortensis*). Annual. Sow seed one month before planting outside. It needs full sun and light, well-drained soil. Blooms midsummer to frost. It's a good accent for mild meats and fish, soups, egg dishes and vegetables. Add the leaves to herb vinegars and salad dressings.

Sweet marjoram (*Origanum majorana*). Annual. Start by seed and plant when the soil is warm. Grows to 30 cm (12 inches). It needs light, well-drained soil and full sun. Blooms August to September.

Vietnamese coriander (*Polygonum odoratum*). Tender perennial. Has a wonderful perfume and is particularly good with poultry and Chinese foods. Overwinter indoors in zone 5 or less.

Perennials and biennials

Perennial and biennial herbs can be left outside over winter, many even in pots, for much longer use. Biennials are grouped with perennials because they perform for two years, and then, if allowed to drop their mature seed, they resprout the following spring and continue the cycle.

Chives (*Allium schoenoprasum*). Perennial. Hardy to zone 3. The Chinese were using chives some 4000 years ago and Marco Polo is reputed to have been the clever young man who introduced them into Europe in the 13th century. Both chives and garlic chives (*A. tuberosum*) need a sunny, well-drained location to perform best. Their mild oniony taste makes them useful in egg and cheese dishes, salads, herb butters and cream cheese, not to mention sour cream.

Fennel (*Foeniculum vulgare*). Perennial or biennial. Hardy to zone 4, this herb grows to 1.2 m (4 feet). Start it from seed; it will reseed itself in following years. It needs well-drained soil and full sun. Bloom time is July to September. It has a pleasant, licorice flavour valuable in Italian dishes and is great for herb vinegar. Ripe seeds are used in sauces, baked goods and tea.

French tarragon (*Artemisia dracunculus* var. *sativa*). Perennial. It's hardy to zone 5, but needs very well-drained soil if it is to survive winter. It grows to 60 cm (2 feet). Divide roots to multiply every three to four years. French tarragon is started by cuttings. (Russian tarragon, which does seed, is inferior for culinary purposes.) Fresh leaves have the strongest flavour. Use them in vegetable and egg dishes, in cream and cheese sauces and with chicken, fish and veal. The leaves make a flavourful vinegar.

Lemon balm (*Melissa officinalis*). Perennial. Hardy to zone 4, it grows to 60 cm (2 feet). It's best started from cuttings or division. Grow it in well-drained soil; it tolerates full sun to shade. Blooms July to September. Nearly mature leaves have the best lemon flavour; use them to make a refreshing tea, add a tang to iced drinks, or flavour salads, chicken, fish, jellies and herb vinegar. Dried leaves make wonderful potpourri.

Lovage (*Levisticum officinale*). Perennial. Hardy to zone 3. Seed in fall or divide in spring. Plant in sun to part shade and give it moist, well-drained soil. It should be divided every two or three years. Blooms June to July. Add the fresh or dried leaves to salads, soups and cheese sauces. The stalks can be used like celery. Established roots can be dug and used in salads. Lovage tea is said to be a diuretic and its taste and aroma are deliciously celerylike.

Mints (*Mentha* species and varieties). Perennial. Hardy to zone 5 and grow up to 60 cm (2 feet). They perform best in full sun and can be started by divisions or cuttings, but with mints it's not starting them that's difficult, it's stopping them! This stuff is invasive, so be warned. You could end up with nightmares of mint leaves chasing you around your yard. There are dozens of kinds with some very interesting flavours, including chocolate. Mint leaves are great in jellies, as garnishes in drinks, or as accents for ice cream or cheesecake. Leaves can be sugared and used as an edible decoration—the list goes on and on. You can even use mints to make a flavoured vinegar.

- Apple mint. Fruity with a slight apple taste. Garnish drinks with it.
- Chocolate mint. The fragrance and taste of chocolate makes it a nice garnish for sweets.
- Orange mint. Citrus flavour that is refreshing in drinks.
- Peppermint. The premier mint for flavouring candies; great in hot or cold tea.
- Pineapple mint. Fruity, pineapple flavour.
- Spearmint. Versatile mint for enhancing meats, salads and soups. Add it to vegetables when cooking.

Oregano (*Origanum vulgare*). Perennial. Hardy to zone 5, it grows up to 60 cm (2 feet). It can be started by cuttings or division. Give this Mediterranean native well-drained soil and full sun. It should be divided every three years. A hardier cousin of marjoram, oregano is used similarly in Middle Eastern and Mediterranean cuisines. Its fragrance makes it great in potpourri, teas and jellies.

Parsley (*Petroselinum crispum*). Biennial. In mild climates (to zone 6), it will overwinter. Grows to 45 cm (18 inches). Sow seed when soil temperatures are over 10°C (50°F). It needs moist, well-drained soil and tolerates full sun to partial shade. Plain-leaf parsley adds a sophisticated flavour to soups and stews. Curled parsley is commonly used as a garnish, but try using it on cooked vegetables and in egg dishes. It freshens the breath (that garnish is meant to be eaten) and is a good diuretic.

Sage (*Salvia officinalis*). Perennial. Hardy to zone 5. It can be started by cuttings or division and needs well-drained soil and full sun. Its decorative blue flowers bloom in June and attract bees. The young leaves are pleasantly bitter and are used in salads, omelettes, breads, meat pies and with poultry.

Thyme (*Thymus vulgaris*). Shrub. Hardy to zone 4. All varieties need well-drained soil and will take sun or part shade. Thyme is great in soups, sauces and vinegars and with fish, poultry and pork. It's used in Italian herb seasonings and as a garnish. Many varieties are popular groundcovers, particularly between paved or stone pathways. *Thymus vulgaris* is most used as a culinary thyme, but *T. citriodorus* (lemon thyme), with its strong lemon fragrance, is also a culinary herb (zone 5), as is *T. herba-barona*, caraway thyme (zone 6).

Winter savoury (*Satureja montana*). Perennial. Hardy to zone 5, it needs a sunny spot with light, well-drained soil. Winter savoury's peppery flavour with just a hint of lemon is often used to season meats, fish, soups, vegetables, eggs and cheese dishes.

Landscaping with Herbs

Many of us don't have the luxury of planting a traditional herb garden. We either set herb plants out in pots or incorporate them into our landscape. Don't overlook the value of ornamental herbs used in combination with other garden plants and shrubs.

Best bets—evergreen perennial herbs

Catnip (*Nepeta cataria*). Zone 3. Grows up to 1 m (3 feet) tall. It needs sandy, well-drained soil and full sun to partial shade. Bears white, purple-spotted flowers from June to August.

Lavender (*Lavandula angustifolia* and varieties). Zone 5. Grows up to 60 cm (2 feet). Start by cuttings or seed. It needs full sun and well-drained soil. Blooms in July and August with lavender-blue spikes.

Lavender cotton (*Santolina chamaecyparissus*). Zone 6. Grows up to 60 cm (2 feet). Start from seed. It needs poor, well-drained soil and full sun. It has gorgeous silvery foliage and yellow flowers in June and July. It's used in formal plantings and knot gardens and the flowers are used in sachets.

Roman chamomile (*Chamaemelum nobile*). Zone 3. Grows to 5 cm (2 inches). Start by seed or division. It needs light dry soil and full sun to partial shade. It's a traditional groundcover between paving stones. Bears white flowers with yellow discs June to August.

Best bets—flowering perennial herbs

Bee balm (*Monarda didyma* hybrids). Zone 4. Grows to 1 m (3 feet). Start by seed or cuttings. It needs rich, moist soil and full sun. Divide every three years. Blooms July and

August with flowers in terminal clusters, surrounded by coloured bracts. Comes in a wide range of red, pink and lavender shades. The dried leaves are used in potpourris.

Betony (*Stachys officinalis*). Zone 4. Grows to 75 cm (30 inches) tall. Start by seed or division. It thrives in moist, well-drained soil and full sun to partial shade. Flowers mid- to late summer. It has been used as a substitute for black teas.

Chives (*Allium schoenoprasum*). Attractive clumps of grasslike green leaves and purple flowers. (See "Culinary herbs" above, for more details.)

Goldenrod (*Solidago odora*). Zone 4. Grows 2 m (6 feet). Start by seed or division. It needs well-drained soil and full sun. Divide every three years. Fresh leaves are a substitute for French tarragon.

Orris (*Iris germanica* var. *florentina*). Zone 4. Start by division. Plant shallowly in well-drained soil and full sun. Blooms May and June with beautiful, light blue, flaglike flowers. The dried and powdered rhizomes are the orris used in perfumes and potpourris.

Purple coneflower (*Echinacea angustifolia*). Zone 3. Grows up to 1.2 m (4 feet) tall. Start by seed or division. It needs fertile, well-drained soil and full sun to light shade. Has become a popular supplement to help boost resistance to infection by aiding the immune system.

Saffron crocus (*Crocus sativus*). Zone 6. Start from bulbs available only in August and early September. It needs light, well-drained

Herbs as houseplants

Growing herbs indoors is not easy to do well but it's rewarding to have your own fresh herbs on those dark winter days. You need an east or north window with good indirect light. In winter, supplement with lighting, preferably fluorescent or sodium halide. Supplementary lighting must be kept on 14 to 16 hours per day during the poor growing conditions of winter.

Grow herbs in small containers to keep them rootbound. Use well-drained potting soil. Water thoroughly, but let the plants dry out between waterings. Feed weekly with 20-20-20 water-soluble fertilizer to keep them green and growing vigorously. If you wish to grow organically, use fish fertilizer. Always water first, then fertilize. Keep room temperatures as low as you can and prune back leggy herbs as they begin to stretch. Letting them become dry, lowering the temperature, providing more light and using low-nitrogen, high-phosphate, high-potash fertilizers all help to keep them more compact. Here are some of the best herbs to grow indoors.

- bay laurel
- blue African basil
- chives
- lemon verbena
- mints
- myrtle
- oregano
- parsley
- rosemary
- sage
- scented geraniums
- thyme
- Vietnamese coriander
- winter savoury

soil and full sun to partial shade. Divide every three years. Flowers are lilac-purple in fall.

Sweet violet (*Viola odorata*). Zone 5. Start by divisions in spring. It needs rich humusy soil and partial shade. Purple, white or yellow flowers.

Best bets—annual and tender perennial herbs

Anise (*Pimpinella anisum*). Blooms in summer with yellowish white flowers that resemble Queen Anne's lace. (See "Culinary herbs" above, for more details.)

Borage (*Borago officinalis*). Grows to 1 m (3 feet). Start by seed after last frost. It needs rich, well-drained soil and full sun. Blooms midsummer with intense blue star-shaped flowers with black anthers.

Myrtle (*Myrtus communis*). Hardy to zone 9, this evergreen shrub can reach 1.5 m (5 feet) tall. It needs moist, well-drained soil and bears white flowers spring to fall. Traditionally used in Europe for wedding bouquets and corsages, as well as potpourris.

Nasturtium (*Tropaeolum majus*). Grows up to 30 cm (12 inches). Spreads or trails up to 2 m (6 feet). Sow after last frost. Blooms summer till frost in poor but well-drained soil. Needs full sun. (See also "Best bets—edible flowers," below.)

Pineapple sage (*Salvia elegans*). Zone 9 perennial. Grows to 80 cm (32 inches). Start by seed or cuttings. It needs rich, well-drained soil and full sun. Blooms June and July. Use the pineapple-scented leaves in drinks. (See also "Best bets—edible flowers," below.)

Pot marigold (*Calendula officinalis*). Grows to 45 cm (18 inches). Sow seed when frost is past. It needs full sun and well-drained soil. Blooms spring through fall with large yellow, cream or orange flowers.

Scented geraniums (*Pelargonium* varieties). Tender perennial. This large group of flowers is generally hardy to zone 9. They can grow up to 1.2 m (4 feet) in size. They are drought-tolerant plants that love full, hot sun and need very well-drained soil. The flowers are used as a garnish.

ips and tidbits

- Garlic should be planted around September 23 in well-drained, raised beds. The best varieties are Russian garlic and elephant garlic. Harvest the following July.
- Mole plant (*Euphorbia lathyris*) is a biennial with nasty roots that deter moles. Plants must be well established to work. Works on gophers, too.
- Mosquito plant (*Pelargonium citrosa* 'Van Leenii') is a tender perennial with a wonderful lemonlike fragrance. To have any effect on insects, you must pick the leaves, crush them and rub them on exposed skin. It's a lot of work, but it's organic, and it works for me.
- Sugar plant (*Stevia rebaudiana*) is a South American native that is actually 300 times sweeter than sugar, all without calories.

Edible Flowers

The idea of munching on marigolds no longer raises quite so many eyebrows as it once did. Indeed, edible flowers have moved from the plates of high-end restaurants into our homes. They look fabulous as a garnish or in a salad, and they spark conversation around the table.

Although any number of flowers would look good on the plate, you have to consider the palate in all this as well, and of course you wouldn't want to serve something that caused any stomach upset for your dinner guests either. When it comes right down to it, even flowers that are safe to eat add much more in their visual appeal than they do in their luscious taste. The larger flowers often end up stuffed with exotic foods, while others are used to add flavours to sauces or salads.

I'll leave the truly exotic to the real experts in this field, but here are some suggestions to get you started.

Best bets—edible flowers

- Anise hyssop. Has a full, spiky pink flowerhead with a flavour that hints at root beer.

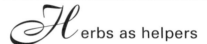 erbs as helpers

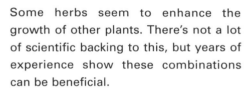

Some herbs seem to enhance the growth of other plants. There's not a lot of scientific backing to this, but years of experience show these combinations can be beneficial.

Herb	Plant it helps
Anise	Coriander
Basil	Peppers and tomatoes
Beebalm	Tomatoes
Borage	Beans, strawberries and tomatoes
Chamomile	Cucumbers, onions and many other herbs
Chervil	Radishes
Chives	Carrots, roses, grapes and tomatoes
Chrysanthemum	Lettuce
Coriander	Anise
Dandelion	Fruit trees
Dead nettle	Potatoes
Dill	Cabbages, onions and lettuce
Garlic	Roses
Horseradish	Potatoes
Hyssop	Grapes and cabbages
Larkspur	Cabbages and beans
Lovage	Beans
Marigold	Potatoes, tomatoes and roses
Mint	Tomatoes and cabbages
Mustards	Fruit trees, grapes and beans
Onion	Beets, cabbages, lettuce and strawberries
Oregano	Beans
Rosemary	Beans
Rue	Figs
Sage	Carrots, cabbages, marjoram, strawberries and tomatoes
Savories	Beans and onions
Tansy	Blackberries, raspberries and roses
Tarragon	Most vegetables
Thyme	Eggplant, tomatoes and potatoes
Yarrow	Most aromatic herbs

- Beans. 'Scarlet Runner' beans are easy to grow and as a bonus, those red flowers are edible and have a nice bean taste.
- Borage. This delightful flower with a cucumber flavour makes an attractive addition to your plate in salads or cooked as greens and in summer drinks. Remove the hairy spikes from the flowers before serving.
- Carnations. The large ones would be quite a mouthful, but the smaller, single-flowered dianthus are more manageable and have a perfumed, cloverlike taste.
- Chives. Those cloverlike, pink blossoms are best used when they first appear, when they have a slight onion flavour. When mature they are tough and tasteless, so get them early.
- Daylilies. The blossoms look good on any plate, but sometimes they have a slight metallic aftertaste.
- Fennel. The dill-like yellow flowers add a delightful touch to stuffed peppers.
- Geraniums. Well known for fragrance, some varieties can work on the plate as well. Try the scented-leaf varieties, such as peppermint. They don't flower very heavily, but the soft pink, white and lavender flowers have a distinct flavour.
- Lavender. The flowers have a wonderful perfume and many people use them to freshen up a linen closet. Most of them taste quite medicinal, but try English lavender (*Lavandula angustifolia*), which is lemony.
- Nasturtiums. These are perhaps the best known of the edible flowers. They taste something like watercress and are quite at home in salads. Use the Whirlybird hybrids—they don't have the black spur usually found in other nasturtiums. Leaves can be a substitute in pepper-restricted diets and the flower buds substituted for capers.
- Pansies. They don't like the heat of summer, but if you have a few in a shaded spot their beautiful blossoms are hard to beat as a garnish. They have a slight floral flavour. I recommend the smaller varieties, such as johnny jump-ups.
- Pineapple sage. The spiky, pink, liatris-like flower stem of pineapple sage can be stripped and the flowers sprinkled over a bed of rice.
- Roses. They're gorgeous, and some roses, such as the rugosa hybrids, are quite tasty. They're also known for their hips, used in making jelly and tea.
- Scotch marigold. Not the tastiest, but quite attractive and safe to eat.
- Squash. Zucchini and many other squash flowers are great to stuff, but 'Butter-blossom' squash is reputed to be the best.

Warning

If you're eating any type of flower, make sure it's one that's been proven safe. Refer to a publication, such as *Poisonous Plants of the United States and Canada,* by John M. Kingsbury, if in doubt. Some dangerous annuals to be avoided are snapdragons, petunias, impatiens, sweet peas and larkspur. Perennials to avoid are primrose, bachelor's buttons, lupine, lily-of-the-valley, *Lobelia cardinalis,* and monkshood. Stay away from the following bulbs: iris, autumn crocus, *Amaryllis belladona* and anemones. Azalea and rhododendron are slightly toxic and should be avoided. Don't use hydrangeas or the exotic weeping wisteria blossoms.

The other consideration is the use of pesticides or herbicides. Make sure you know the source of the flowers and can ensure they haven't been treated with toxic spray.

Bedding Plants

Colour. In a word, that's why it's important to have annuals, or bedding plants, in the garden. They give a splash of brilliance that heightens the impact of any bed. Starting in late spring and carrying through to fall, they take ownership of many gardens, dominating the landscape with colour and form. Today's gardeners rely heavily on perennials, but there is still a need for long-blooming annuals that carry a continuous colour theme through the growing season.

Annuals have changed a great deal over the past few years with extensive, worldwide breeding programs. Our old favourites are sporting many new colours, but many also have exciting foliage and a more compact growth habit for today's smaller gardens. I try to visit the North American Annual Trials each year and I never fail to come away inspired. In the area of colour, particular attention is being paid to analogous colour mixes—colour combinations based on various shades of two or three colours next to each other on the colour wheel, such as red and orange. Two good examples are 'Cranberry Punch' impatiens, a blend of cranberry, salmon picotee and coral, and the vibrant 'Sunrise Mix', a blend of apricot, coral, orange, salmon and white. They are both sensational, and all because of the careful use of analogous colour (plus white for a dash of contrast).

Breeders are also widening the scope of many annuals. For example, they have bred tuberous begonias, coleus and impatiens to be more sun tolerant and geraniums, celosia and nicotiana to be more shade tolerant.

Today we have many types of rainproof petunias and more sun-tolerant pansies and violas. Triploid marigolds that do not go to seed and therefore maintain their vigour are outperforming the old-fashioned French marigolds. Snapdragons branch out all by themselves and keep blooming, while new, long-lasting, compact varieties of the forgotten linarias have brought their early blooming form back from oblivion.

It pays to purchase these new varieties because they have been bred to bloom longer under stressful conditions, such as extreme heat, cold and wet. We all have our old favourites, but sticking exclusively to them is like betting on the same racehorse for 10 years. That horse is now older, slower, more susceptible to chronic ailments and less able to function well in adverse weather. Try the many new varieties; if gardeners don't embrace them, the hard work and field trials of the breeders will be out the window, and you will be missing out on the benefits.

Working with colour and shape

When I speak to gardening groups I often say the true test of a great garden is to take a lawn chair and sit down in front of a particular bed and drink it in for a while. If you can sit there for more than five minutes and really enjoy what you see, and notice more and more depth as time goes by, then that bed has really got it. On the other hand, if you've seen everything there is to see in a nanosecond, you've got some work cut out for you.

Analogous colour plantings help build depth by using two or three colours next to each other on the colour wheel as well as their tints and hues, such as blue, blue-violet and violet. The results can truly be amazing. Monochromatic colour means using all the tints and shades of one colour. It's fabulous, but in our own gardens it often falls short of the depth analogous colour can provide. Complementary colours are those opposite each other on the colour wheel, such as green and red, orange and blue, or yellow and violet. These colours tend to contrast with each other for instant attention, but often fail the five-minute test.

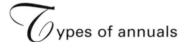ypes of annuals

The word *annual* refers to plants that complete their entire life cycle within one year or less. They are usually started as seeds or transplants in spring, flower for a given time from spring through fall, then die out with the first heavy frost. Some plants we refer to as annuals are actually perennials in milder climate zones; they just don't survive our harsher winters. Annuals are often classified as hardy, half-hardy or tender. This gives you a guide when considering their suitability for your growing environment.

Hardy annuals can tolerate light frosts at the beginning and end of their growing season and can often be seeded when the soil temperatures are very low. Many of these plants will reseed themselves for the next season. California and Icelandic poppies, alyssum, snapdragons and nasturtiums are some examples of good self-seeders.

Half-hardy annuals are quite tolerant of long periods of wet, cold weather, but will succumb to hard frosts. Seeds of these plants must be sown after the last frost, but will germinate in cooler temperatures. Stocks, petunias, dianthus, calendulas, lobelia and scabiosa are some examples in this category.

Tender annuals are generally native to warm climates. In their native environment they are perennial, but in our colder climate they are grown as annuals. They need warm soil to germinate and perform best in long, warm summers. It's usually better to start these plants indoors or purchase transplants to bring them up to speed more quickly in your garden. Some tender bedding plants are also less tolerant of extreme heat. Some examples are begonias, impatiens, geraniums, salvia and zinnias.

If you mix deep blue browallia with deep purple heliotrope and toss in some lavender and salmon impatiens, it's stunning. At the famous Rogers Garden Center at Newport Beach, California, I saw 'Tapien' verbena that covered a sloping bank in purple, light blue, soft pink and lavender; the effect was breathtaking. In our own gardens, a bank of pink mophead hydrangeas was left alone and the soil pH changed with the heavy rains. The result—a tapestry of pink, lavender, blue and purple all shading together. It was a favourite spot for about two months.

Blending any analogous colours (and you don't have to be exacting here) produces wonderful effects. The nice thing with annuals is, you can keep changing the colour each season and each year until you find the most pleasing combinations.

When considering colour, use warm and cool colours to advantage. In the cool months of the year, challenge winter by using warm colours such as yellow, orange and red in your garden. These colours have high visual energy and psychologically make you feel warmer—like long underwear for the eyes. They also create an illusion of advancing toward you, making large areas seem smaller.

When selecting your winter pansies or violas for example, use more yellow, orange and red shades. I like to use soft yellows with maroon-reds. Choose striking colour combinations, and plant them in large enough groupings to create impact. Mass plantings along your driveway, sidewalks, steps and around feature trees really make a difference. Use other warm foliage plants—compact golden conifers, broadleaf evergreens, evergreen perennials and berry plants—to further heighten the effect.

When it's hot, cool colours are assets in your garden. Icy pinks, blues and greens make small spaces seem larger and give an illusion of coolness. Pink-toned impatiens and fibrous begonias, blue browallias and lobelia and green-tinted coleus are as good as sipping iced tea for making you feel a lot cooler on a hot day.

Many folks prefer to use a blend of all colours. This can succeed in creating a somewhat happy effect, but it can look a bit untidy. If you use warm and cool colours together, the cool ones tend to be overwhelmed by their brazen counterparts. To really improve the look of your landscape, use complementary, monochromatic or—my personal favourite—analogous colouring, and you will see a dramatic difference.

White, silver or chartreuse are great accent colours to highlight any plantings. All three colours are light and add a nice tint to your colour schemes. The effect can be remarkable, often adding excitement to an otherwise quiet combination.

One commonly overlooked topic that makes a huge difference to the look of a landscape is the use of edging or borders. Using one or two plants that work together as an edging adds continuity and flow—often the missing ingredient in tying things together. Hardy edging plants can be in place very early and will remain viable late into the year, when there is not much blooming. Borders can blend plants together that somehow never seem to tie in. Unless it's a very formal setting, use smooth sweeping curves to help create a flow.

Spacing depends on the plant and how big it grows, but I recommend putting annuals too close together rather than too far apart. Think of how well plants blend together in baskets and planters when they're snugged up tight, and translate this into the garden, where it looks just as good. Most of us want a fast, good-looking planting and bedding plants set miles apart take too long to fill in. Novice gardeners tend not to group plants, but larger displays of fewer plants look better than smaller displays of many plants. Good-sized plantings of seven to nine plants create a very professional look.

Best bets—annuals for easy start from seed

Direct-seed these right where you want them to grow in your garden. They will produce long-blooming, easy-care plants. All on this list are easy to grow and have that old-fashioned look. (See chapter 1, "Seeds and Seedlings," for information about direct seeding.)

Name	Colour	Size (height x width)	Sun/shade	Hardy, half-hardy or tender
African daisy (*Dimorphotheca*)	yellow-orange tones	30 cm (12 in.)	sun	half-hardy
Calendula or English marigold	yellow, orange shades	30 cm x 45 cm (12 in. x 18 in.)	sun	hardy
California poppy (*Escholtzia californica*)	bright yellow, orange-red tones	30 cm (12 in.)	sun	hardy
China aster	all colours	various	sun	hardy
Cosmos	pink, white, rose tones	dwarf: 45 cm (18 in.) tall: 90 cm (3 ft.)	sun to part shade	half-hardy
Livingstone daisy (*Mesembryanthemum*)	pink, salmon tones	15 cm (6 in.)	sun	hardy
Nasturtium	orange, yellow, red, cream	30-60 cm (1-2 ft.)	sun to part shade	hardy
Portulaca	bright yellow, orange, red, white, pink	13 cm (5 in.)	sun	hardy
Strawflower (*Helichrysum bracteatum*)	red, yellow, white tones, pink	dwarf: 30 cm (12 in.) tall: 90 cm (3 ft.)	sun	hardy
Sweet peas	pink, red, white tones	dwarf: 30 cm (12 in.) tall: 90 cm (3 ft.)	sun to part shade	hardy
Zinnias	red, yellow, orange, pink, white	dwarf: 30 cm (12 in.) tall: 90 cm (3 ft.)	hot sun	half-hardy annual

Best bets — annuals for edging

All of the plants in the chart below make outstanding edging plants with minimal care once established. To create an effective border, be meticulous about planting in a smooth-flowing line. Staggered plantings can help disguise crooked lines. Keep plants far enough away from lawns and pathways so that maintenance doesn't become a problem with invasive plants.

Name	Colour	Size (height x width)	Sun/shade	Hardy, half-hardy or tender
Ageratum	blue, white	10-15 cm x 20 cm (4-6 in. x 8 in.)	sun to part shade	tender
Alyssum	white, pink and purple	8-10 cm x 30-45 cm (3-4 in. x 12-18 in.)	sun to part shade	hardy
Dusty miller	silver	15-20 cm x 20-25 cm (6-8 in. x 8-10 in.)	sun to part shade	hardy
Dwarf feverfew 'Golden Moss'	white flower with chartreuse foliage	15-20 cm x 20 cm (6-8 in. x 8 in.)	sun to part shade	hardy
Dwarf marigold	orange, yellow	15-20 cm x 20 cm (6-8 in. x 8 in.)	sun	tender
Fibrous begonias	white, pink, red, rose	15-20 cm x 20 cm (6-8 in. x 8 in.)	sun to part shade	tender
Impatiens	all colours	15 cm x 60 cm (6 in. x 24 in.)	part shade to shade	tender
Lobelia	blue, white, pink, rose	10-15 cm x 30 cm (4-6 in. x 12 in.)	shade, part shade to sun	half-hardy
Veronica andersonii variegata	variegated foliage	20-25 cm x 30 cm (8-10 in. x 12 in.)	sun to part shade	hardy
Violas	all colours	15-20 cm x 20 cm (6-8 in. x 8 in.)	sun to part shade	biennial

Best bets — annuals for cut flowers

Annuals provide a wonderful source of cut flowers from late spring to fall. Virtually any long-stemmed annual will do for cutting, but a few stand out because of their longevity, fragrance or interesting colour or shape. Many of the following varieties are used commercially in the floral industry, but they are easy to grow and stand up well in the garden. You won't go wrong if you try them.

Carnations (*Dianthus caryophyllus*). 'Chabaud's Giant Double Improved Mix' has the widest colour selection of any garden carnation variety. It bears large, almost commercial quality blooms on 45-cm (18-inch) stems. Bloom June to August.

China asters (*Callistephus chinensis***).** The Matsumoto types have 60-cm (24-inch) stems that hold a wide range of colourful 5-cm (2-inch) semidouble blossoms with contrasting yellow centres. Bloom July to September.

Common statice or sea lavender (*Limonium sinuata***) Pacific Series.** I like the colour range of these vibrant papery flower clusters. They tie cut flower bouquets together as a filler and also look great by themselves. They dry easily as well. Bloom July to October.

Cosmos (*Cosmos bipinnatus***) Sensation Series.** Fast-growing 90-cm (3-foot) plants produce large 9-cm (3½-inch) blooms. They are reliable performers in all weather and very heat tolerant. Good colour range—pinks, rose, white, picotee red, and rose and white bicolours. Bloom July to October.

Larkspur (*Consolida ambigua***) Earlibird Series.** These have long stems with fully double flowers in a wide range of pastels as well as deep blues and carmine shades. They grow to 50 cm (20 inches) and are 2 to 3 weeks earlier than most other varieties. Bloom July to September.

Marguerite daisies (*Chrysanthemum frutescens* **or** *Argeranthemus frutescens***).** This is actually a tender perennial that will sneak through mild winters, but don't count on it. It will take early frosts and provide some of the earliest cut flowers. The compact strains produce stems long enough to cut, but it is the range of colours from pastel yellow to vibrant pinks that give them new prominence. The semidouble varieties bloom all through the summer for a long cutting season. Bloom May to October.

***Rudbeckia* 'Indian Summer'.** This award-winning variety has enormous 15- to 23-cm (6- to 9-inch) single, deep golden yellow blossoms with contrasting dark brown centres. They are very wind tolerant and look fabulous in a vase by themselves. Don't ignore all the other rudbeckia varieties. Bloom July to October.

***Scabiosa caucasica* 'Finest Mix'.** One of the renowned cut flowers in pastel shades from rose and lavender to pure white and coral. They grow to 1 m (3 feet) with elegant, double, anemone-type blooms. Bloom July to August.

Snapdragons (*Antirrhinum majus***).** Snapdragons are an old favourite, but the newer compact varieties are much improved. Liberty Series is my pick of all the snaps for a wide colour range and good spikes of evenly spread florets. They are very sturdy, grow only 45 to 70 cm (18 to 28 inches) and branch nicely. Bloom May to July.

Heat-tolerant bedding plants

Even heat-tolerant annuals need good soil and some time to become settled in before they can tolerate the full intensity of the sun. In areas of fast-changing weather patterns, where days of cloud and rain give way to scorching sunlight, these plants will need extra care and moisture until they adapt.

- African daisy
- cockscomb
- common salvia
- cosmos
- fibrous-rooted begonias
- gazanias
- geraniums
- petunias
- portulaca
- rudbeckia
- spider flower
- triploid marigolds
- zinnias

◀ Brassicas underplanted with red-leafed lettuce make great companions.

Photo: National Garden Bureau

▼ Hybrid broccoli is best to harvest before the blossoms appear.

Photo: National Garden Bureau

I love to munch on the leaves of *Stevia rebaudiana,* the sugar plant, for a sugar high.

▼ I grow my herbs close together for minimal weeding and continual harvesting. I like to use some ornamental herbs like artemisia to add a little accent.

◄ Old-fashioned nasturtiums are easy to grow for continuous colour and as a source of edible flowers. A little lime helps keep those pesky aphids away.

Photo: National Garden Bureau

▼ Continual summer-flowering shrubs such as this one ensure bright spots of colour in your garden throughout the season.

Photo: Island Sun

▲ Shade basket of nonstop begonias, impatiens, fibrous begonias, lobelia, variegated vinca and silver nettle.

► A tropical vine such as this one can provide fabulous colour all summer and into pumpkin time. Don't forget to bring them inside for winter.

◄ By adding a trellis to planter baskets, you can achieve a whole new dimension of colour and effect.

▼ Supertunias with bacopa make a great combination. I'm learning to keep colours and combinations simple and effective.

▲ Black-eyed susan is one of the longest-blooming hardy perennials to enhance late summer and fall.

▲ Blue fescue and Japanese blood grass create wonderful garden accents, especially when blended with other perennials like *Artemisia* 'Silvermound' and *Centranthus ruber*.

◄ Sedum 'Autumn Joy' is an attractive old standby not only for fall colour but for its icy blue-green foliage.

▶ Floribunda roses tend to be compact and floriferous, for almost continuous summer colour.

▼ David Austin roses combine the old-fashioned look and fragrance of antique roses with the continuous-blooming habit of modern roses.

Stocks (*Matthiola incana* '**Brilliant Double Purpose**'). These are fragrant old-fashioned favourites that love either shade or a bit of morning sun. They produce 60-cm (2-foot) stalks with columns of delightful pastel flowers. They dislike hot weather, so plant them early. Bloom May to July.

Sweet peas (*Lathyrus odoratus* '**Cuthbertson Floribunda**'). These have extra-long stems with 6 to 7 large blooms per stem. It's a very heat-tolerant variety providing lots of flowers even in hot weather. Excellent colour range. Bloom May to August.

Zinnias (*Zinnia elegans*). The cut-and-come-again hybrids are free-flowering and fully double. The 8-cm (3-inch) blooms flower and reflower incessantly to provide an abundance of colour in the heat of summer and into fall. They grow to 70 cm (28 inches) and come in a wide range of zinnia colours.

Best bets—bedding plants for a shady spot

Planting in a shady location can be a learning experience. Virtually no annual will perform well in heavy, dark shade. Shady spots should ideally be fairly light and bright at least part of the day, but be careful if you have a shady spot that gets full sun between 11 a.m. and 3 p.m., when it's at the peak of intensity. The rule is the heavier the shade, the taller the plants will grow in search of light. Early and late sun is fine with most shade-loving annuals, and in fact will encourage stronger plants. Some shade-lovers, such as fibrous begonias, will tolerate a good deal of sun.

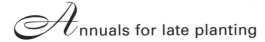nnuals for late planting

If you've been snoozing away the spring, and it's into June, it's still not too late to get some annuals in the ground. Many old fashioned annuals are available as seed or bedding plants and will help colour up your garden in short order. They grow quickly when it's hot and add a fresh new look and long-lasting colour just when the garden is beginning to look a little tired in the heat.

African daisy, or star-of-the-veldt (*Dimorphotheca* **hybrids**). Forms a mat of brilliant orange, cream and yellow flowers that bloom until frost.

Brown-eyed susans (*Rudbeckia* **hybrids**). These plants have been overlooked for too long. They love the heat and provide great colour for the midsummer garden. Try 'Rustic Mix' or 'Marmalade.'

Cosmos (*Cosmos bipinnatus*). Shoots up quickly and blooms with huge, pastel flowers.

Evening scented stock (*Matthiola longipetala*). Blooms just two months after sowing and continues for a couple of months.

Garden balsam (*Impatiens balsamina*). An old favourite that performs well in areas with shade or morning-only sun.

Painted tongue (*Salpiglossis sinuata*). Related to the petunia, it produces ornate blossoms on 60-cm (24-inch) stems.

Spider flower (*Cleome hassleriana*). Makes a nice display with the long-blooming white blossoms of 'Helen Campbell' or the soft pink 'Rose Queen'.

Browallia (*Browallia speciosa*) 'Major' and the Bell Series. These relatively unknown shade-loving annuals bear masses of starlike, soft blue or white flowers all summer. They grow to 30 to 35 cm (12 to 14 inches) and are great for baskets and planters.

Coleus (*Solenostemon scutellarioides*). (See "Best bets—foliage plants" below.)

Fibrous-rooted begonias (*Begonia semperflorens*), Cultorum hybrids. These 15- to 20-cm (6- to 8-inch) sun-loving begonias tolerate light shade quite well. The waxy foliage is drought tolerant once established, making them ideal for dry shady spots and under trees. They are continuous-blooming. The green-foliage varieties are best in shade.

Flowering tobacco, or nicotiana (*Nicotiana alata*), compact varieties. These sun lovers actually perform equally well in light shade. Magnificent new salmon, apple blossom and cherry blossom colours are wonderful to work with. Grow to 25 to 30 cm (10 to 12 inches).

Fuchsia (*Fuchsia* hybrids). Upright forms with small flowers have been forgotten in favour of the trailing varieties until recently. Now there is a resurgence of the small multiflora varieties, which produce masses of self-cleaning blossoms all summer and well into fall. 'Tom Thumb' and 'June Bride' are great garden varieties.

Garden balsam (*Impatiens balsamina*), double varieties. One of the unknown great garden plants with huge columns of double flowers that last all summer. Slugs seem to avoid it. Grows to 30 cm (12 inches).

Heliotrope (*Heliotropium arborescens*). There are many new varieties in the market today, and many have a delightful perfume that pervades your entire garden or patio. Plant white or soft pink flowers around the deep purples for an effective contrast. White heliotrope looks best in shade. Grow to 25 to 35 cm (10 to 14 inches).

Impatiens (*Impatiens walleriana*). The number-one shade-loving annual with long-lasting blooms in a wide range of colours. Plants grow to 20 to 25 cm (8 to 10 inches) and quickly spread to smother the ground in colour.

Lobelia (*Lobelia erinus*). In morning sun or shade situations, the compact clumping forms create wonderful borders, edgings and underplantings. Trailing forms are old standbys for planters and baskets. Colours vary from rich purple blues to white, pink and soft blues. Grow to 8 to 10 cm (3 to 4 inches).

Monkey flower, or mimulus (*Mimulus* hybrids), double variety. These plants love light shade and cool weather. They grow to 25 to 30 cm (10 to 12 inches). They can go in the garden early for a great show and bounce back in the cool fall with more colour. Compact varieties like the Mystic Series are free-flowering and spreading. The older varieties, like 'Yellow Velvet' and 'Royal Velvet', have huge flowers that stand up to cool, wet spring weather.

Tuberous begonias (*Begonia tuberhybrida* hybrids). The NonStop Series is the best, with 9-cm (3½-inch) fully double blossoms in a wide range of colours that just don't quit. Purchase either starter seedlings or 10-cm (4-inch) potted plants for instant colour. Plants grow to 30 cm (12 inches).

Best bets—cool-weather annuals and biennials

After a long, grey winter, we're often overzealous when it comes to getting some colour into our gardens, especially when the first warm sunny days arrive. Long, wet, rainy spells and frosty evenings can quickly put an end to any annual that's not able to withstand a few degrees of frost. This is where hardy annuals and biennials play such an important role.

But be warned: even these plants need to be well acclimatized to outside conditions before you can set them out. Cold frames—a simple wood frame with clear polyethylene stretched over top—are the ideal way to harden off these plants. Place them in a cold frame for about four days before leaving them out without protection. (See chapter 1, "Seeds and Seedlings," for more information.)

Failing this, setting them outside, out of direct sun and wind, for a few days will help. The cooler the outside temperature, the longer the plants will take to acclimatize. If they are not properly acclimatized, they could take weeks to recover, or even die. A few days of careful preparation will pay off with early colour and vitality.

Here are some of my picks for cold-hardy colour—no more than 10°C (18°F) of frost. They are all plants I feel comfortable recommending for a dose of early spring colour. They're fairly tough and won't let you down

Growing geraniums

Without a doubt, these are some of the most popular bedding plants in our gardens, and with good reason. There has been a lot of improvement in geraniums in recent years and if you have been growing the same varieties for years, I suggest you look into some of the new ones. Growing from seed is a good bet— it's far less expensive, rewarding in its own right and gives you access to the latest, improved varieties.

To start from seed, don't buy the cheapest seed, buy the newer hybrid varieties, which will perform better. I suggest the Orbit, Ringo or Elite series. Start them in Jiffy 9 or Jiffy 7 peat pellets—one seed per pellet. Keep them moist and humid by covering them with a clear plastic bag, and most importantly, maintain a bottom heat of 21°C (70°F). The seeds will sprout in two to three days, and when they do, give them plenty of light and good air circulation. A fluorescent fixture, with one grow light and one cool white tube about a metre (3 feet) over the plants, is ideal. Give them about 16 hours of light a day.

To encourage short, compact growth, keep them continually pinched back, dry and under good light. Don't make the all-too-common mistake of letting them get too leggy. Keep up with the pinching, either by nipping out the centre growth bud on each stem, or cutting the whole stem back with a sharp knife just above a leaf joint or node. Geraniums have an amazing ability to send out new growth, even out of hard, woody, old stems. Prune them back at any time, but try not to let them get away on you.

Geraniums like it on the dry side. If they are rootbound, the water will disappear rapidly, allowing them to dry out quickly. Allow them to dry out between waterings. If you grow them in large pots with lots of damp soil, they will either get root rot or become tall and leggy. They will also bloom much later. They don't need much in the way of nutrients. Start them off with 20-20-20 and then switch to a low-nitrogen, high-phosphorus fertilizer, such as 15-30-15, 10-52-10 or 0-15-14. These fertilizers will encourage earlier blooming and will help keep the plants compact. (For more on geraniums, see "Best bets—geraniums" in chapter 5.)

Growing bedding plants from cuttings

You don't have to say goodbye forever to all of your bedding plants at the onset of the first heavy frosts. These "slips" will give you winter colour as well as a source of new plants in the spring. True annuals are not suitable for propagation in this manner, but I recommend it for geraniums, fibrous begonias, impatiens, marguerites, and many of the trailing plants from your hanging baskets.

If you follow some basic rules, as well as a few of my trade secrets, you should have success in propagating annuals. You have to remember your cutting is making a big change in growing conditions—from the extremes of outdoors to a warm, dry indoor environment.

1. One key to success is to keep your rooting area as humid as possible. The best way to achieve this is to use a standard 25- by 50-cm (10- by 20-inch) flat, lined with a "cell pack" insert, and entirely enclosed with a polyethylene tent. This mini-greenhouse can be created by covering the flat with a dry-cleaning bag.

2. Next, prepare your rooting medium. A good rooting medium provides good drainage along with excellent moisture retention. In the spring and summer, use straight mica peat, but in the fall I recommend coarse sand or a mixture made of equal parts perlite, sand and peat. Moisten the medium thoroughly.

3. Select the strongest and most healthy shoots as cutting material. Use a clean, sharp knife and cut the stem just below a set of leaves, giving you a piece 8 to 10 cm (3 to 4 inches) long. Make your cut at a node because this is where new roots will form more quickly.

4. To reduce stress on the cutting, trim off about half the leaves.

5. Dip the cut end of each slip in rooting hormone powder and place it firmly in the moist planting medium.

6. Once the cuttings are in place, never let the planting medium dry out. Mist the cuttings several times a day with warm water. A systemic fungicide, used as recommended on the label, will help control fungus and disease problems. See your favourite garden centre for advice.

7. Grow lights will speed up the process and prevent your cuttings from becoming too leggy. Otherwise, grow your cuttings beside a window where they will have lots of indirect light, and keep the plants reasonably cool, in the 15°C (60°F) range.

Cuttings of impatiens, begonias and marguerites, should root in seven to ten days. Geraniums may take up to two weeks. Tug gently to see if they are firmly rooted. When they are, remove the plastic tent, but continue misting until they show no signs of wilting. Give the top of each slip a pinch to encourage branching and feed once a week, alternating with 20-20-20 and 5-15-5. Leave the rooted cuttings in their cells as long as you can, then transplant into an 8-cm (3-inch) pot, using a high-quality potting soil. Keep them rootbound until spring, when you can place them in the garden.

when the weather turns sour (unless it's a truly disastrous spell, and then all bets are off!). They should keep you satisfied until the weather warms up enough for some of the more tender plants.

Dusty miller (*Senecio cineraria*). Fall or spring. Dusty miller doesn't have significant flowers, but it has terrific foliage. Plant chrysanthemums or pansies with it for colour. (See "Best bets—foliage plants" below.)

English daisy (*Bellis perennis*) Rusher or Galaxy Series. Fall or spring. These are both winter- and spring-flowering, with weather tolerant, small, button-type flowers that bloom persistently in moderate winters. Come in white, red or pink.

Marguerites (*Chrysanthemum frutescens* or *Argeranthemus frutescens*). Spring. Try the dwarf European varieties. I've had good luck with the yellow-flowered 'Proven Winner'. 'Mini-white' has wonderful blue foliage, and 'Dwarf White' is good as well. Use them along borders, in planters or in mass plantings.

Monkey flower, or mimulus (*Mimulus* hybrids). Spring. Mimulus really steals the show come June. (See "Best best for a shady spot," earlier in chapter.)

Ornamental kale and cabbage (*Brassica oleracea*). Fall. (See "Best bets—foliage plants" below.)

Medium- and small-flowered pansies (*Viola × wittrockiana*). Fall or spring. Small- and medium-sized flowers are by far the best performers in all weather. Large-flowered varieties are a wonderful novelty in ideal, sunny but cool weather.

Pinks (*Dianthus* hybrids) Ideal Series. Spring. Growing to only 15 to 20 cm (6 to 8 inches), the single-flowered pinks are a real treat. They're at their best in small groupings or containers. The new varieties are free-flowering, more compact and longer lasting in hot weather. The single-blossom varieties tolerate heavy rain.

Snapdragons (*Antirrhinum majus*). Spring. They'll take frost and some will survive a mild coastal winter. 'Sonnet' and 'Liberty' are mid-sized and strong. For a dwarf, try 'Floral Showers'. (See also the listing under "Best bets—annuals for cut flowers" above.)

Stocks (*Matthiola*) East Lothian Series. Spring. This old European strain usually overwinters in cold frames with a little extra protection. It tolerates early frosts and can usually be set out by early March. It loves early morning sun situations and needs dry feet. The columnar flower in pink and lavender tones is highly perfumed. Plant them in shady spots if you want them to last longer.

Toadflax (*Linaria maroccana*). Spring. These colourful little annuals resemble miniature snapdragons and are best sown where they are to bloom. A March sowing produces April flowers that thrive in cool wet weather. They tend to burn off in hot sun but are a welcome treat early in the season. The Fantasy Series is more compact and longer flowering.

Violas, or mini-violas (*Viola cornuta*). Fall or spring. These are the hardiest of all cool-loving biennials. They will take severe frost and still bounce back. Violas are stronger flowering than their cousins, the large-flowered pansies, especially in adverse weather conditions.

Best bets—foliage plants

We often think of bedding plants in terms of colourful flowers, but many are better known for their dramatic foliage. These plants can be used as focal points or dramatic backdrop plantings.

Castor bean plant (*Ricinus communis*). Both the green 'Sanguineus', which grows to 2.4 m (8 feet) and the burgundy 'Zanzibarensis', which grows to 4.5 m (15 feet) have spectacular foliage that makes them ideal in the centre of large beds or as a screen. We've used them for years now and they never fail to impress visitors.

Coleus (*Solenostemon scutellarioides*). These wonderful foliage plants look fabulous the moment they are planted. The Wizard Series is basal-branching and fills out all summer without any pinching, unlike the old-time varieties. The chartreuse varieties are the most vibrant from a distance. They do best in dry spots where slugs are not as prevalent, but if you set them out later, when it's hot, slugs are not as much of a problem. They will tolerate a great deal of sun; the new Solar Series from Florida is the most sun tolerant of any variety to date, and they are truly spectacular.

Dracaena palms (*Cordyline australis*). These plants are started from seed one year in advance of planting in the ground. Green or bronze foliage is traditional, but a series from England has great new colours. 'Red Star' has brilliant burgundy-bronze foliage on a very compact plant. 'Purple Tower' has wide, purplish burgundy leaves with a dramatic flare. 'Sundance' has green edges with distinct vibrant pink centres. 'Coffee Cream' has dark burgundy leaves with creamy stripes along the sides.

Dusty miller (*Senecio cineraria*). This is one of the most important accent plants in any garden. The serrated foliage of 'Silver Dust' has made it a favourite. My favourite is 'Cirrus', the most spectacular silver variety, which tolerates rain and frost best. The feathery variety known as 'Silver Lace' is a different species—*Tanacetum ptarmiciflorum*. It is elegant, but I find it does not stand up as well in damp climates.

Flowering maple (*Abutilon* varieties). These South American natives are evergreen tender perennials. In England, virtually every public garden seems to use white and green variegated abutilon as a focal point in show beds. They are grown as a bush or trained into a standard tree form—a dynamic display. *A. megapotamicum* 'Variegatum' is a graceful trailing form. *A. pictum* 'Thompsonii' is a compact upright form with creamy yellow foliage.

Ornamental kale and cabbage (*Brassica oleracea*). A good choice for fall plantings. Ornamental kale is ruffled; cabbage has a more rounded leaf. Those started later produce smaller heads, which tolerate winter cold (up to zone 6) and rain far better. The open, lacy "peacock" varieties are the most weather tolerant.

Snow on the mountains (*Euphorbia marginata*). This amazing plant, with its variegated green and white leaves, grows almost anywhere in hot sun or partial shade. It's tolerant of heat or drought and poor soils. It fits nicely in either formal or informal settings. Grows to about 60 cm (2 feet).

Hanging Baskets and Planters

Hanging baskets and containers overflowing with colourful flowers and graceful, trailing vines can transform any outdoor living area. Some plants also scent the air, a bonus on warm summer evenings. That's the good news. The challenge is to create or purchase a container that will not only look great initially, but continue to grow and perform well all summer long. After 30 years of creating baskets, I have learned a few things only too well. Perhaps most important: don't rush the season at the first sign of nice weather.

There are four distinct seasons in areas with a moderate climate. The first is early spring, after the hard frosts are over, when cool-tolerant baskets can be planted. For the best performance from heat-loving container plants, don't put containers out until the weather warms up and stays warm. Mother's Day seems to be a time when containers are first set out, but another two or three weeks makes a tremendous difference. Wait until it really warms up and your plants will thank you by growing vigorously and providing a fabulous display well into the fall.

In the fall, replace tired, scruffy-looking summer displays with late-blooming hardy plantings. With some protection, baskets of fall perennials and violas will last through winter and continue into spring in zone 6 or above.

For the holidays, a few evergreen boughs, cones and interesting winter branches carefully tucked into moss baskets can create a delightful Christmas planting to accent your December decor.

The ABCs of Baskets and Planters

It's a good idea to be extra choosy when selecting varieties for the special growing conditions and look of hanging baskets and containers. They should be fast-growing and vigorous to make a full-looking basket. The plants must have good foliage colour, or at least a flower that shows up from a distance. I expect a long flowering period under adverse conditions, and I expect plants to tolerate a bit of neglect, in case they're overlooked now and then. Avoid messy plants that are constantly shedding flowers unless you really like cleaning your baskets up every day. Also look for pest-tolerant species; anything that is susceptible to mildew or insects, such as spider mites, should not be used. It's a tough set of rules, but if you want to get results rather than regrets from your containers, it's worth it.

Before you choose a single plant, there are some important decisions to make. First and most important is the size of the container. Make no mistake—the larger it is, the better. Small containers put plants under too much stress on those hot, windy days. A 30-cm (12-inch) size is the minimum, and bigger is even better. Plastic or wood baskets are just fine, but if you really want a showpiece, wire baskets lined with moss or cocoa fibre and wood byproducts are fabulous because you can also plant the sides.

Volume of soil is important for root development and moisture retention, so you'll need depth as well as diameter in your containers. For faster results, however, use baskets that are lower and wider. The plants are able to fill up the sides more quickly, because once they quickly cover the shallow sides, the basket immediately looks great and you need fewer plants to create the look you're after.

Soil is critical to success; don't use bargain-basement specials or garden soil. A good-quality sterilized soil will drain well and retain moisture. Some folks prefer soils with moisture-retaining "hydro-gels." These tiny gels absorb excess moisture, which can then be accessed by plant roots, like a "spare tank" of water, when they need it. These soils expand by up to 30 percent when watered, so leave room for expansion in the container.

Some of the more successful commercial basket growers are using chips of rock wool (mineral wool) blended into the soil, and are having superb results. Use what works for you, but please don't use garden soils, and don't mix fertilizers into the soil—better results are achieved if you top-dress.

When you fill the baskets or containers, don't pack the soil too tightly. Tap it up and down gently to settle it in. To set plants in,

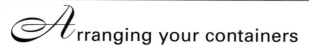

Arranging your containers

When placing your plants, decide which is the front and which is the back. Most plants have an aspect that is most attractive. I like to group upright plants in the centre. Pinch them back lightly so they will bush out. Place trailing plants near the edge of the container. Pinch long, leggy trailers back to two or three sets of leaves, so that a fuller plant will develop.

Some plants are better in a container by themselves or with other similar types. For instance, licorice vine and *Bidens aurea* tend to overtake everything else. Combine plants with similar growth habits, or be ready to prune vigorously and often. Balance your containers by allowing for equal numbers of vines and colour on opposite sides.

For the most dramatic effect in colours, use analogous colours—tints and hues of colours next to each other on the colour wheel. Use plenty of silver, white and chartreuse for accent. Cool colours work best in summer and warm colours are most appreciated in winter. Most of us use a potpourri of all colours for a polychromatic effect. It's nice, but believe me, once you get the hang of analogous colouring, you will be hooked!

take out a handful of soil the size of the rootball and gently set the plant in place. Add a little soil to get things level again and cover the roots, and that's it. Do not push the plant into the soil, and make sure when all the plants are in that the soil level is just under the rim of the container, allowing the first watering to settle the soil line down a few centimetres (an inch or two). This will create a rim so that water can pool before it percolates through the soil.

Planting is only the beginning—it's your growing technique that will show your true gardening colours. Proper watering is essential. When you water, especially when containers are newly planted, use warm water and water thoroughly. Let's not mince words: soak the bejeepers out of it. If it's a wire basket, soak the sides too. Wait until the basket becomes slightly dry before you water again. How do you tell? Give it a lift. Dry baskets are somewhat light; wet ones, heavy. Watering may seem simple, but it's surprising how many plantings fail to perform to their full potential due to improper watering techniques. Follow these simple steps and you will be on your way to becoming a master of hanging basket and container gardening. Drip irrigation systems work well, but the same principle applies.

Every time you water you are giving the plants a good drink, but you're also leaching out nutrients. Containers without fertilizer will go downhill quickly. Fertilize them at least once a week (twice is even better) with water-soluble 20-20-20 fertilizer. In addition, use a single application of slow-release granular fertilizer on top of the soil. This double dose will make all the difference in the world.

As the plants mature, reduce the nitrogen by switching to 15-30-15, which will result in more blossoms. Near the end of the season, change to a 0-10-10 formula or 0-12-12, which will boost blooms. An organic fish fertilizer with a similar formulation will add bacteria to the soil, a bonus.

The last and ongoing chore is to constantly remove dead or spent blossoms. Be extra vigilant with geraniums: blossoms that fall on other plants can burn the foliage. A little tidying up makes a significant difference not only in the container's appearance, but also in the plants' ability to keep on blooming. Spent flowers go to seed and sap a plant's energies, so off with their heads! One last grooming tip: hyperactive vines should be systematically pruned back to maintain the integrity of the display.

Best bets—early spring containers

When the worst of winter is over, usually in late March in zone 6 and above, cool-loving plants that tolerate a light frost can be set out for a splash of early colour. If it suddenly turns very cold, move the containers to a protected spot.

Australian violet (*Viola hederacea*). Spreads quickly over the edges of the container and gives a very natural look. Tiny

---

Container first aid

If your container dries out or is over-watered, underfertilized or otherwise stressed, it's possible to bring it back from the brink. Follow these tips and you'll be surprised how quickly it bounces back.

- Soak the soil and roots with a fungicide if there are rot problems caused by overwatering.
- Allow the plants to dry out more between waterings, but mist the foliage.
- Cut back all unhappy plants.
- Use diluted 20-20-20 as a foliar spray every day to encourage new growth.

blue and white flowers bloom all spring into summer.

Bacopa (*Sutera cordata*). (See "Best bets—trailers," this chapter.)

Golden creeping jenny (*Lysimachia nummularia* 'Aurea'). A warmer colour for cool weather.

Marguerite 'Sugar Baby' (*Chrysanthemum frutescens* or *Argyranthemum frutescens*). A new, very compact, white-flowering variety that really does fit well with other plants in containers.

Monkey flower, or mimulus (*Mimulus × hybridus*). The Calypso and Magic hybrids with small, yellow, orange, red and pink flowers are a must.

***Nemesia denticulata* 'Confetti'.** (See "Best bets—trailers," this chapter.)

Nepeta (*Glechoma hederacea* 'Variegata'). An attractive and fast-growing trailer with round, scalloped, green and white leaves.

Pansies (*Viola × wittrockiana*). If you want to use pansies, at least use the medium-sized types. The big guys will let you down.

Silver nettle vine, or yellow archangel (*Lamium galeobdolon*). One of the best for rapid growth.

Variegated ivy (*Hedera helix* varieties). The variegated forms add rich texture to any basket.

Variegated periwinkle (*Vinca major* 'Variegata'). Produces beautiful blue flowers in late spring.

Violas (*Viola* species) small and mid-sized varieties. These are actually tiny pansies.

Best bets—trailers

Following are some of my favourite trailing plants, but I reserve the right to add different ones at any given moment. Every year new cultivars show up in the stores, so pick the ones that strike your fancy but be open to new ones that come along.

Bacopa (*Sutera cordata*). This is a superior basket plant with its great trailing habit and ability to combine with almost any other plant. The mauve and pink varieties don't perform as well, but 'Snowflake' and 'Snowstorm' tolerate most weather and flower continuously with multitudes of tiny white flowers.

***Bidens aurea*.** If you're looking for a fast-growing, indestructible trailing plant, then look no further. Marigolds look great, but can fade out under intense sun. *Bidens* has lacy

Growing container plants from seed

If you have a small greenhouse or are good at starting seeds, you can save a lot of money by buying seed packets and growing your own "basket stuffers," as everyone in the industry calls them. Some are easy to start, like creeping zinnia, petunias, lobelia, gazanias and signet marigolds, often referred to as "Tagetes." If you have lots of containers, propagating your own plants will make a big difference to your budget.

Lead time is necessary to have your plants available when the containers are ready to go outside, and seed catalogues will have the approximate starting times and germinating techniques you need. (For information on growing from seed and transplanting, see chapter 1.) Don't forget to keep good records for next year.

foliage and single yellow blossoms that just don't quit. The more compact variety, 'Gold Marie', will blend in and enhance instead of overtaking any container. You can start it from seed or cuttings. If you want to cover the world in about 30 seconds, use the original *Bidens,* 'Golden Goddess'.

Coleus (*Solenostemon* hybrids). Many of the newer dwarf, basal-branching types such as 'Wizard Mixed' are much improved over older types and add fullness to baskets. 'Golden' is perhaps the finest of their many colours. It doesn't trail very far, but with a few pinches can be coaxed along.

Creeping zinnia (*Sanvitalia procumbens*). This very durable plant actually seems to prefer abuse. It's a semivine that winds its way around other flowers. The orange variety, 'Mandarin', isn't as rugged as 'Gold Braid', but the colour is nice.

Fanflower (*Scaevola aemula* 'Blue Wonder'). This has been around for some time and it's a great plant, but it needs both heat and sun for optimum output. It's a funny vine that sort of flops over the edge of the basket or container, but throws out a steady stream of lavender-blue flowers that look quite exquisite. Use a bit of *Artemisia* 'Silver Brocade' to highlight the blue tones. There are also newer fanflowers in white and lavender-pink, with both large and small flowers and a slightly nicer drooping habit.

Licorice vine (*Helichrysum petiolare*). If you purchase a jungle machete, you can use this wild and fast-growing trailer. Its small, velvety, silver leaves always provide a great contrast but 'Rondello', silver with a dark blotch, is sensational. Chartreuse 'Limelight' is absolutely magic. Prune hard and often when they're young; if not kept in check, they'll soon dominate the container.

Lobelia (*Lobelia erinus*). This is an old favourite as a basket stuffer, but it's hard to beat and comes in blues, whites, reds, mauves and purples. Try 'Trailing Sapphire' with its striking blue accents. 'Blue Cascade' has light blue flowers, and I really like the 'Colour Cascade' mix. They don't perform well in hot, dry locations, so keep them in a shadier spot.

Lotus vine. Performs well in sun and shade, and as long as it's watered it will look great right until the end of the season. 'Amazon Apricot' has salmon-apricot blooms. 'Amazon Sunset' has vibrant orange blossoms. 'Berthelotii Scarlet' is a lacy, silver vine with red flowers in late summer. 'Maculata Gold Flash' is lacy with thick silver foliage and gold flowers.

***Lysimachia congestiflora* 'Golden Globes'.** It's a vigorous plant with dull green leaves and long-lasting yellow flowers. It blooms better in sun than shade and takes some time off between blooming. 'Outback Sunset' has striking variegated leaves.

***Nemesia denticulata* 'Confetti'.** This South African native is a tender perennial that blooms extravagantly from April through October. It's more of a flopper than a true trailer. Prune it back when young and enjoy its masses of tiny, pink, snapdragonlike flowers. I'm also delighted with its close cousin 'Bluebird', with its dark lavender-blue flowers. They make a great couple.

Slipperflower (*Calceolaria* 'Sunshine'). It takes a while to start this from seed, but it's worth it. It seldom needs pinching and cascades nicely. Dozens of tiny flowers last from early in the season until fall. Prefers cooler weather and performs well in shade.

Swan River daisy (*Brachycome* 'Ultra'). This plant's lacy foliage and large lavender-blue, daisylike flowers are great for softening the look of any basket. It's at home in sun or shade. Yellow, white and pink colours are available in both jumbo- and small-flowered varieties.

Trailing fibrous begonias (*Begonia* 'Avalanche'). This low-maintenance plant is at the top of my list. It's a drooper rather than a trailer. It performs equally well in sun or shade and blooms almost nonstop from early in the year until long after others have packed it in. 'Pink Avalanche' has a better perfume than the white variety. In cooler weather, the foliage provides a nice bronzy contrast.

Trailing petunias (*Petunia × hybrida*). Suddenly out of nowhere came a series of new trailing petunias mostly grown from cuttings rather than seed. Supertunia, Surfinia and Cascadia are three slightly different kinds developed and promoted by different companies. They are all similar in their incredible growth rate, up to 1.5 m (5 feet) and beyond and producing masses of flowers that bloom incessantly all summer in the most intense heat. They are tender perennials (zone 9) and will, in fact, start like a geranium in spring if cut back and kept dormant in a frost-free area over winter.

A graceful arching habit and rapid growth make them far superior to any previous petunia in a basket or planter situation. They are, for the most part, self-cleaning and quite rain tolerant. If they do suffer mildew problems, cut them back and they'll bounce back very well.

All varieties need heavy feeding all through their growing and blooming stages. As well as slow-release fertilizer, they should be fed at least twice a week, first with 20-20-20, then 15-30-15 for best results. The only other tip is to prune them back constantly when they are young plants to get them very bushy before they are allowed to trail over the edge of the basket or planter.

One variety introduced from Japan, which is actually started from seed, is the amazing 'Purple Wave'. It was an All-American selection and, in my opinion, is the best of the lot, both with its vigour and flowering habit. The magenta flowers create a carpet of colour. 'Pink Wave', introduced a year later, is not quite up to the standard set by 'Purple Wave', being a little more upright in its habit, but a whole series of new 'Waves' is on the way.

There has been such a profusion of trailing petunias introduced so quickly that not all of them provide top performance in gardens, baskets and planters. Remember, they can be used as a summer groundcover as well. Each of the many varieties performs differently in varying summer climates but, after a few years of trials, here are my picks. By the way, you may want to use them by themselves in a basket, because they tend to overpower everything else.

- 'Blue Vein'. Medium-sized blossoms of lavender with purple throat and purple veining. Soft fragrance.

Trailers for moderate shade

If you have difficulty finding time to maintain baskets and planters, but love the effect they provide, using colourful vines and trailers can give you a good display with minimal maintenance. These will all tolerate shade.

- asparagus fern (*Asparagus densiflorus* Sprengeri Group).
- *Centradenia inaequilateralis*. Rich bronze foliage, light purple flowers; a wonderful foliage contrast.
- creeping jenny (*Lysimachia nummularia*).
- licorice vine (*Helichrysum petiolare*) 'Limelight'.
- silver nettle vine (*Lamium galeobdolon*). All varieties.
- variegated catmint (*Plectranthus fors*).
- variegated dead nettle (*Lamium maculatum*). All varieties.
- variegated ivies (*Hedera*) 'Variegata'.
- variegated periwinkle (*Vinca minor*) 'Variegata'.

- 'Brilliant Mini-Pink'. Small, vibrant pink blossoms with a light throat; incredibly prolific.
- 'Pink Vein'. Medium-sized, soft lavender-pink blossoms with magenta veining.
- 'Purple Sunspot'. A medium-sized violet-magenta with a black centre. It's a great trailer with a compact habit.
- 'Sun Lace'. Slightly ruffled pink blossoms that shimmer in sunlight.
- 'Sun Snow'. Purple veins in huge white blossoms and an attractive trailing habit.
- 'Surfinia Blue'. Deep purple, medium-sized blossoms that are a real eye-opener when blended with pink.
- 'Surfinia Purple'. The original and still the best with its magenta flowers, neat trailing habit and massive flower production.
- 'Surfinia White'. Pure white, large blossoms that just keep coming.
- 'Sweet Victory'. A large-flowering, hot pink variety.
- Some interesting additions are the tiny-blossomed trailers called 'Million Bells' and 'Liricashower'. The blossoms are less than 2.5 cm (1 inch) in diameter, but they are profusely produced in a very nice trailing habit. They can stand alone in smaller baskets, but also play an important role in mixed baskets because they are the first petunias that will not overpower other plants. They come in bright rose-pink and a clear mid-blue.

Verbena Tapien. I like trailing verbena, especially during hot summers, because it really thrives in the heat. Unfortunately, a few damp days or too much water on the leaves can bring on the dreaded mildew, which spreads to other plants. At last, there is a mildew-resistant series, called Tapien, which comes in pink, purple, blue and lavender, with more colours on the way. Its lacy foliage and continuous blooming are welcome improvements over the older verbenas. The Temari Series has enormous flowers in a good range of colours with more traditional but mildew-resistant foliage.

Best bets—geraniums

Ivy, or trailing, geraniums (*Pelargonium peltatum* hybrids). These have become the standard by which other basket plants are measured. They stand up well in a wide variety of weather conditions and keep on blooming and looking good with minimal care. Here are some of the best.

- Balcon Series. For larger baskets and planters. Similar to the mini-cascades but larger single flowers on longer stems. Red, lilac and pink.

Best mite-free plants for hot sun

Fanflower (*Scaevola aemula* 'Blue Wonder'). Blue, fan-shaped flowers. Loves heat.

Mini-cascade geraniums. Smaller plant that fits in well with other basket plants and never quits flowering, unless steady rainfall causes root rot in late summer.

Nemesia denticulata 'Confetti'. Small, fragrant snapdragonlike flowers bloom in profusion all summer. Needs pruning.

Petunia 'Million Bells'. Small-flowered, heavy-blooming petunias that blend beautifully in baskets without overpowering other plants.

Strawflower (*Helichrysum bracteatum*). Incredibly heavy-blooming yellow strawflower that just doesn't stop.

Swan River daisy (*Brachycome* species). All love the heat and have attractive foliage and continuous-blooming daisylike flowers.

- Blizzard Series. Large, continuous-blooming varieties, very rain tolerant. 'Red Blizzard' is a true red and 'White Blizzard' is a good-performing white. 'Blue Blizzard' is lavender, and there's a 'Pink Blizzard' as well.
- Fischer varieties of trailing geraniums. 'Amethyst' is the best dark lavender semi-double. 'Beach' is a true dark red. 'Butterfly' is the best light lavender. 'Luna' is white. 'Sybil Holmes' is the best bright double pink, blooming like a miniature rose. 'Mexicana' is a somewhat leggy red and white.
- Galleria Series. Very heat tolerant with nice zonal patterns on the leaf. One of the earliest to bloom. 'Pink Passion' is a clear pink, 'Ruby Red' is the darkest red and 'Sunrise' is an orange-red. Although trailing, they are a little more upright.
- Guillou Series. 'Bernado' is coral-red. 'Blanche Roche' is white. 'Charade' is deep lavender. 'St. Malo' is dark red. 'Vinco' is hot pink, 'Wico' is salmon and 'Acapulco' is pink.
- 'Mini-Cascade'. Simply the best for smaller baskets and window boxes; it produces a solid mass of colour spring through fall. Weather tolerant and comes in pink, lilac and red single blossoms. All full, hot sun.

Zonal geraniums (*Pelargonium* × *hortorum*). The upright, compact Fischer Series of zonal geraniums are noted for their short bushy habit and strong flowering ability. They are my choice for upright basket geraniums, especially if you want to combine them with other plants. For dark red, try 'Tango'; deep red, 'Samba'; hot pink, 'Blues' or 'Casanova'; rich salmon, 'Schöne Helena'; white, 'Lotus'; purple-red, 'Magic'; magenta, 'Tango Violet'; and deep purple, 'Kardino'.

ew Guinea impatiens

These heat-loving large-flowered *Impatiens* hybrids are now taking a good portion of the basket and planter market by storm. It is important that they be put out only when the weather warms up and stays warm, because they will almost surely go backward in cool weather, especially if they are overwatered.

Once established, in typical warm summer weather, they are remarkable. Not only do they have large flowers, they are self-cleaning, bloom continuously until frost and have a very attractive mounded or domed growth habit.

These are very heat-tolerant plants, but they do not like full sun all day. Morning or later afternoon sun is fine, but they do stress out, especially in baskets and planters, if left to the sun's mercy from 11 a.m. till 3 p.m. I have found that those with darker foliage tolerate more sun than the green or variegated types, once they are well established in garden beds. There are many good series available today. Try the Australian Rainbow Hawkeye Series, which is most sun tolerant, the Pure Beauty Series and the Paradise Series.

Try planting bacopa around the perimeter of a container planted with New Guinea impatiens for a very beautiful effect.

Best bets—fuchsias

Fuchsias are truly a powerhouse of colour, especially in moderate summer climates. They need partial shade and in hot weather they love to be misted with water during the day. Make sure the leaves are dry in the evening to avoid disease problems.

One of the key factors for success with fuchsias is to water the baskets thoroughly, but let them dry out well between waterings. If the soil remains wet for too long the roots become waterlogged and rot away rather quickly. If in doubt, it's better to water less and mist the foliage more.

Whitefly and spider mites are the two major pest problems for fuchsias. An application of the insecticide recommended by the dispenser at your garden centre (I have had good success with methoxychlor) two or three times on a weekly basis will control whitefly. Spray early in the morning before they become active. Yellow sticky cards are a good control if your whitefly problem is minimal.

Misting with a fine mist on the underside of the leaves will control mites. If that fails, use a mite spray, two or three times in succession, about one week apart.

Fertilize fuchsias weekly with soluble 20-20-20 to start, then 15-30-15 when they bloom. Add a slow-release fertilizer on top of the soil once the plants have settled in.

Everyone likes the big double varieties, but the semidouble and single varieties outperform the big guys. Here are my picks for the best fuchsias for containers. (If you don't see your favourites on the list, it's probably because they don't flower continuously like the following ones.) All are singles, except for 'Lena' and 'Streamliner', which are semidouble. *Note:* The outer flower parts are called the sepals; the inner flower parts, petals that make up the corolla, are sometimes called the skirt.

Trailing	Corolla colour	Sepal colour
'Cascade'	carmine	crimson
'Jack Shahan'	rose-pink	rose-pink
'Julie Horton'	deep purple	soft pink
'Lena'	deep purple	soft pink
'Marinka'	crimson	crimson
'Miss California'	white	rose
'Red Spider'	rose	crimson
'Streamliner'	rose-red	crimson
'Tinker Bell'	pink-veined white	pale pink

Upright	Corolla colour	Sepal colour
'Black Prince'	deep purple	red
'Dollar Princess'	purple	cerise
'Gartenmeister Bonstedt'	salmon-orange	salmon-orange
'Jingle Bells'	white	red
'June Bride'	rose	light rose
'Mary'	bright crimson	bright crimson
'Tom Thumb'	purple	rose
'Tom West'	carmine	carmine
'Traudchen Bonstedt'	light orange	salmon-pink

If you still want those big, double, mouth-watering fuchsias, here are my picks.

Double	Corolla colour	Sepal colour
'Bicentennial'	marbled coral-rose	light coral
'Blue Eyes'	lavender-blue	red
'Blush of Dawn'	lavender-blue	white
'Dark Envy'	violet-blue	deep red
'Lisa'	lavender	rose-pink
'Personality'	marbled purple-rose	rose-red
'Pink Marshmallow'	light pink	light pink
'Swingtime'	white	pale red

Best bets—fragrance

No basket or container is complete without fragrance, and heliotrope is tops. Its white, blue or purple blossoms perfume the air.

A close second is white alyssum. White is an important colour in containers, and alyssum has a gently cascading habit that softens the edges of planters and baskets.

If you have a sunny location, try fragrant herbs in a basket or container. Creeping rosemary, variegated thymes, such as lemon thyme or the softer silver thyme, and golden marjoram are all suitable.

Alyssum 'Snow Crystal' (*Labularia maritima*). This white variety is one of the best. Sun to partial shade.

Carnation 'Flame' (*Dianthus caryophyllus*). Very fragrant, deep rosy pink flowers begin to flower in June and bloom in profusion all summer. Full sun to light shade. Try it in a basket by itself or plant it with bacopa.

Heliotrope (*Heliotropium arborescens*). Medium blue, dark purple and white. Sun or partial shade. It's tough and will flower right up until frost with minimal care. Feed it with fish fertilizer for good blooms. 'Fragrant Delight' is the darkest purple and the most compact, although not the most perfumed. Many varieties of heliotrope have been around for years with no specific cultivar names. The white is also very fragrant, although the colour tends to be a bit muddy.

Nemesia denticulata **'Confetti'**. Truly one of the outstanding plants for baskets and planters in full sun. Thousands of tiny snapdragonlike pink blossoms open in April and continue till frost. They need to be pruned continually and fed at least weekly. Their soft perfume is strongest in the hot sun.

Trailing petunia (*Petunia × hybrida*) Surfinia 'Blue Vein'. Small lavender blossoms with magenta veining have a soft, enticing perfume.

Fall and winter containers

Winter doesn't have to mark the end of the season for container growing, especially in milder areas. There are some interesting plants that can make our late fall and winter gardens look vibrant and colourful between snowstorms and stay that way until spring. Use the plants suggested here, but combine them with winter-flowering daisies, winter pansies and violas to add some floral colour.

- Many winter plants offer foliage colour with unique forms and shapes, and they all increase their colour intensity with cooler temperatures for an even more exciting display. Focal points are the key to success with winter plants and two of the finest are the new tricolour bush dogwood, *Cornus sanguinea* 'Midwinter Fire', with its vibrant orange, yellow and red stems, and the

ℱlowering plants for moderate shade

Many container plants blend nicely with others for wonderful combinations. All those listed below make wonderful long-lasting container plants for moderate (not heavy) shade. Many plants also look great by themselves in a container. Experiment.

- Blue browallia with silver licorice vine.
- Blue lobelia with silver licorice vine.
- 'Illumination' tuberous begonias with white bacopa and pink 'Tapien' verbena.
- Impatiens with white bacopa.
- Pink- and yellow-toned coleus with blue trailing lobelia.
- 'Pink Avalanche' fibrous begonias with blue lobelia.
- Pink impatiens with soft blue or pink 'Tapien' verbena.
- Trailing fuchsias with blue trailing lobelia and 'Rondello' licorice vine.
- Tuberous begonias with variegated ivy.
- White or pink impatiens with blue trailing lobelia.

contorted red and yellow willows, *Salix tortuosa* 'Golden Curls' and 'Red Curls'. These plants light up any basket or container.

- Ajugas, with over 40 species, are on top of my list, especially the following four cultivars, all hardy to zone 5. 'Mahogany' is of German origin with shiny, deep rich mahogany leaves which set it apart from all others. Its flat and rather tight growth habit makes it ideal for container work. 'Burgundy Glow' is an unusual variegated form with rich pink new growth which deepens in cool weather and matures to white, green and burgundy. It's a little more trailing by nature and is ideal in winter hanging baskets. 'Rainbow' is a distinctive, uneven blend of yellow, orange and deep red burgundy—a great fall combination. It is also compact—ideal for planters. 'Silver Beauty' has wonderful grey-green leaves edged in white. They all need good air circulation to prevent mildew in winter.

- A most unusual plant is a beautiful, spiky, silver-blue helichrysum, *Helichrysum thianschanicum* 'Icicles', hardy to zone 5. It looks tender, but apparently it's a toughie, and it looks for all the world like silver icicles. It adds pizzazz to any planting.

- Hardy evergreen grasses are simply made for winter. My favourite is *Carex hachijoensis* 'Evergold'. It has a gently flopping habit and cascades softly over a basket or planter. *Acorus gramineus* 'Ogon' is similar, although more upright, and works well as a focal point. Growing in somewhat of a fan shape, it adds a little spike and height to planters, and it contrasts nicely with virtually anything. *Carex buchananii* is a focal point with its threadlike bronze stems, and its cousin, *Carex* 'Frosted Curls', with long silver-blue threadlike stems, is a great trailer. Both of these will tolerate zone 6.

- A golden oldie is the yellow form of creeping jenny, *Lysimachia nummularia* 'Goldilocks'. In winter it becomes a little more chartreuse in colour, making it a superb accent plant for the darker-toned ajugas.

- One of my favourite old perennials, *Euphorbia amygdaloides* 'Purpurea', has black-green foliage that turns burgundy in cool weather. It makes a good focal point, and the chartreuse-yellow flowers in spring are always a treat. 'Chameleon' is another colourful variety with bronze foliage.

- The more I see the ornamental salvias, the more I like them. The three winter "toners" are *Salvia officinalis* Purpurascens Group, 'Icterina' and 'Tricolor'. All three deepen in colour as night temperatures dip. Purple salvia is nice to use with anything silver, while the tricolour variety adds spark anywhere it goes. My favourite of the three is the golden yellow variety, simple yet elegant. With a little pinching, they are low-spreading and gorgeous.

- For a silver accent, lamiums are quite outstanding. An old and often forgotten variety called 'Herman's Pride' looks for all the world like the old-fashioned indoor aluminum plant with its small silver leaves netted with green veins. This German native flops more than it trails, but it really accents other plants. In spring it has golden nettlelike flowers, and is a good four-season plant. For shady areas, *Lamium maculatum* 'Aureum', with its golden foliage, is fabulous.

I suggest using these plants in zone 6 and above because when you plant in containers, you lose one zone. And don't forget: extra protection is needed during those really cold spells.

Winter hanging baskets

Making some hanging baskets for the holiday season can really boost the spirit.

When my fall baskets start to look bedraggled, I know it's time to start adding winter greens. It takes some experimentation to get it right, but it's possible to make a basket that stands up to winter winds and looks great for weeks at a time. The secret is to use a 30- or 35-cm (12- to 14-inch) wire basket frame lined with lots of lush green moss, and packed with moist soil or peat. Next, cut fresh greens about 30 cm (12 inches) long from your trees and shrubs, and firmly set the ends into the basket so the tops hang down gracefully. Make sure the greens are secure enough to withstand heavy winds. The basket goes together much more effectively if you hang it up, or set it on a high table, so you can see how it will look from a hanging position. Don't cut the branches too long; you are aiming for a balanced, graceful appearance.

The type of greens you use will vary according to availability: those on the west coast will use cedar and cypress, and those in colder areas will use spruce and pine. Try mixing a couple of types together.

Once you have the look you're after, the fun really begins. I use burlap or raffia bows in the centre, with long tails hanging over the side. Berries look wonderful tucked in here and there, but be sure to make them look like they belong. Pine cones can be wired into place so they hang down a bit. White-tipped cones show up better. If you're after a flashier appearance, use gold, silver or coloured ribbons rather than the raffia. Match the colours with whatever else you have in your holiday decor.

These baskets don't need much care, other than some misting of the soil and greens in sunny or windy environments to keep them moist. The moss and soil should keep the greenery fresh, as well as provide some weight so branches don't go flying in winter winds. After the holidays are over, you can lengthen the enjoyment by taking out some of the Christmas touches and putting in a small wooden bird feeder, creating an attractive bird-feeding station.

Perennials

This enormous category of plants includes alpines, herbs, bulbs, bog plants and grasses as well as the many familiar flowering perennials. This broad-ranging group grows in a wide variety of settings, ranging from arid desert to shady rainforest, so always try to match the ideal growing conditions for each perennial to your particular site. Some are very low maintenance, while others require regular pruning, staking and dividing. Most are herbaceous and die down each fall, while others have resilient foliage that looks attractive all year in moderate climates (zone 6 and above).

Perennials are nonwoody herbaceous plants that live longer than two years. Most varieties will last for many years and continue to look better each year with proper care and eventual dividing to renew the plants. Others, like *Meconopsis betonicifolia* and *Primula vialii,* are rather short-lived and need replacing after a few years—but they're worth it.

The trick to using perennials is combining them with other plants for an effective year-round display that changes with each season. Even winter should provide an opportunity to create interest. One of the challenges is to learn not only the flowering times of perennials, but how to use the beauty of their foliage to keep your beds looking attractive longer. Many gardeners use perennials in beds by themselves, which is effective when the plants are in their prime flowering time, but leaves the space underused the rest of the year. This is where flowering bulbs, broadleaf evergreens and colourful, compact conifers come into play. Winter-flowering heathers bloom for almost five months—at a time when perennials are at their weakest. Many evergreen ornamental grasses, like variegated sedge or sweet flag, look great all winter and serve as effective covers for herbaceous perennials in winter, while accenting them in spring. Early-blooming winter aconites, snowdrops, crocus, grape hyacinths and glory-of-the-snow can be planted to pop up around perennials for an early spring kick-off to the emerging perennial foliage. It's exciting to experiment with these combinations and watch the interplay through the seasons.

Creative gardeners have found many ways to use perennials in the landscape. In addition to herbaceous borders and island beds, consider underplanting trees, foundation plantings, driveway and walkway edgings and my favourite, container plantings. There is a very special sophisticated look created when perennials are used in combination with annuals and small shrubs in containers. In many cases, particularly in zone 6 or higher, containers with the right combinations will continue not only to look good, but will change in appearance with each season. A little tip: remember that you lose at least one zone of hardiness when you contain perennials in unprotected pots all winter.

Care and feeding

Remember that all perennials need at least some attention. Pruning back tall-growers like delphiniums and lupines after they finish blooming only improves their appearance and will often result in a second flowering. Tall, large-flowered perennials such as peonies will require support rings to prevent rain and wind from knocking the heavy blossoms down.

Mulching each fall or early spring with well-rotted manures is essential for good growth. One of the drawbacks of mulching is that it can encourage a lot of vegetative growth. This can cause overcrowding and make taller plants more susceptible to wind and rain damage. Once the growth has started in earnest I always apply 4-10-10 or a similar low-nitrogen, high-phosphorus and high-potash fertilizer to keep the plants more compact and increase the size and number of blossoms.

It's a good idea to mulch tender perennials with fir or hemlock bark or sawdust for extra winter protection. Well-drained soils are essential, especially over winter. Over the years I've lost more perennials to root rot than to extremely cold weather.

Organic controls for pests and diseases are always the best way to go, but there are many effective and safe (if used properly) home garden pesticides that can help situations that are out of control. Long periods of wet weather may create fungal problems, like mildew and black spot, that can quickly ruin a wonderful scabiosa or phlox plant. To identify the problem and find a solution, contact the department of agriculture, university extension department or botanical garden in your region.

If your plants are healthy, bloom well and are not overpowering other plants, let them be. But usually after three or more years, most perennials simply get too massive for their location, and they start to decline in flower size or look unsightly. This is the time to divide them. Early fall, as the plant begins to die back, is usually the best time. Each plant has a unique way of being divided, but the basic rule of thumb is to lift out the whole clump and break it apart with two garden forks, a sharp, square-nose shovel or one heck of a big knife. Divide it into good-sized blooming clumps and replant them as soon as possible so new roots can develop for next year.

Not all perennials can be divided in this way. Many, like candytuft (*Iberis*) and rockcress (*Arabis*), are low, spreading plants that originate from a single stem. Pruning them back hard is one way to tidy up their appear-

> ### Crowding out the weeds
> A trend that has developed with today's busy lifestyles is planting perennials and their companions closer together. This creates an instant, bold-looking bed, and the bonus is that the weeds have a hard time finding anywhere to grow. We have tried this in a 120-m (400-foot) border along our roadside, and I've yet to see a weed patch grab a major foothold in this area.

ance, but in many cases, side stems root into the soil and with a little careful digging you can actually separate clumps of rooted stems to start new plants.

If you're simply moving a perennial from one spot to another, do it during the dormant season. This is any time from late fall until new growth begins to appear in the spring. Dig the plant up with a good clump of soil to protect the roots.

Perennials should be an attractive garden component throughout the entire growing season. When they begin to look untidy, cut back the unsightly parts and feed them with a fertilizer low in nitrogen and high in phosphorus and potash to pick them up. If all the leaves are diseased or sun-scalded, most plants can be pruned back to ground level. You'll be surprised how quickly they bounce back and bloom again.

Best bets—perennials for a shady spot

One of the questions I hear most often is how to create colour and life in shady areas. Very dark areas will by their nature only permit the successful use of certain plants like ferns and foamflowers. We continually struggle with this problem in our gardens, and where possible we climb up trees and do selective pruning to allow more light. What a difference it makes in the number of plants we can then use to create some wonderful effects.

When I recommend these shade plants, it's for areas that have light to moderate shade, not heavy shade. In some situations, a shady area receives full sun for part of the day. That's not a problem in the morning or evening, but shade-loving perennials prefer to be out of the sun when it's at its peak, between 11 a.m. and 3 p.m.

Hostas, of course, are the undisputed masters of the shade garden. A detailed listing of hostas follows this section.

Bleeding heart (*Dicentra formosa* 'Luxuriant'). Zone 2. Red, white or pink. Blooms all summer. Grows to 45 cm (18 inches). Although the flowers of 'Luxuriant' are not quite as spectacular as the old-fashioned *D. spectabilis,* they are beautiful and have a long bloom time. Maidenhairlike foliage looks good all summer. It's not bad in containers, and a little morning sun tends to make it bloom longer. Divide every three years to maintain vigour.

Creeping forget-me-not (*Omphalodes verna*). Zone 6. Grows to 13 cm (5 inches). Spreads rapidly to create an amazing display of tiny blue flowers from April to May even in heavy shade. It's a spectacular underplanting for star magnolias.

Foamflower, or fringe cups (*Tiarella cordifolia*). Zone 4. Grows to 30 to 60 cm (12 to 24 inches). Shade to part sun. Most varieties of this B.C. native have masses of foamy white and pink flower spikes in spring. They're great for attracting bees to your garden, and have attractive black-blotched foliage. They are evergreen in mild areas, and the red and

eonies

I'm often asked about peonies that have stopped blooming. Dividing may help. In late fall, dig near the crown and take a look at next year's shoots. There should be two types: thin shoots are the leaf stems and thicker ones are the buds. To ensure your peonies keep producing those desirable fat shoots, make sure to divide the plants when they get huge.

Peonies prefer well-drained soil and benefit from a dose of 4-10-10 fertilizer in early spring when the growth begins and again after blooming is finished.

bronze winter foliage makes them valuable year-round plants. Look at 'Pink Bouquet', 'Inkblot', 'Tiger Stripe' and 'Ninja'.

Goatsbeard (*Aruncus dioicus*). Zone 4. Creamy white flowers in summer. Grows from 1.2 to 1.8 m (4 to 6 feet). Part shade. *Aruncus dioicus* looks for all the world like a tall white astilbe and is a great backdrop for other shade-loving companions. It loves dry feet. The full foliage colour is a rich bronze and the flowers make a great addition to any cut bouquet.

Golden ray (*Ligularia dentata*). Zone 4. Yellow flowers bloom in summer. Grows to 1.2 m (4 feet). The foliage is impressive, but the tall flower spikes are nothing less than spectacular. These are big background plants that need moisture. My favourite is 'Rocket', which has long, thin bottlebrush spires of bright, lemon-yellow flowers on contrasting black stems.

Hardy fuchsia (*Fuchsia magellanica* 'Riccartonii'). Zone 6. Red or violet. Blooms all summer. Grows 1.8 to 2.4 m (6 to 8 feet). Many fuchsia varieties are being promoted as hardy, but they vary greatly in their ability to withstand cold winters. Good drainage, some mulching and a location out of cold winter winds give fuchsias the best chance for survival. 'Riccartonii' is an old reliable performer and provides a timely boost of colour in late summer and fall.

Jacob's ladder (*Polemonium caeruleum* 'Brise d'Anjou'). Zone 5. Blue blooms in summer. Grows to 90 cm (36 inches). Part sun to shade. The blue flowers, although nice, play second fiddle to the magnificent soft yellow and green variegated foliage of this plant. It adds a spark to any shady spot.

Lungwort (*Pulmonaria* varieties). Zone 4. Pink, blue, red and white flowers in spring. Grows to 30 cm (12 inches). In mild areas they are evergreen and the silvery foliage of many new varieties is spectacular. Some favourites are 'Excalibur', 'Spilled Milk', 'David Ward', 'White Wings', 'Longiflora', 'Berries and Cream', 'Victoria Brooch', 'Polar Splash', 'Lewis Palmer', 'Little Star' and 'Raspberry Splash'. Blend them with scillas and primulas in a woodland setting. Some varieties are mildew prone in wet areas.

Yellow corydalis (*Corydalis lutea*). Zone 6. Grows to 20 cm (8 inches). This yellow corydalis has been my longtime favourite. The maidenhairlike foliage and abundant, continuous flowers all summer make it a winner when combined with its self-seeding and spreading habits.

Yellow loosestrife (*Lysimachia punctata*). Zone 4. Yellow flowers with variegated foliage bloom in summer. Grows to 90 cm (36 inches). Part sun to shade. 'Alexander' has variegated leaves and with a bit of sun in your shady spot, the variegation will be more intense. Its yellow flowers are most vibrant in June, but the attractive foliage will brighten the garden the rest of the year. It opens with a pink tinge, then becomes a distinct creamy white.

Hostas

These popular plants deserve a section of their own because there are so many variations in terms of leaf size, colour and flowers. Over 1000 varieties are available, from miniatures to huge 1.5-m (5-foot) specimens. Some will tolerate hot sun, while others prefer heavy shade. We use masses of them at the gardens to spice up shady corners or give a splash of colour under trees.

Iris

Gardeners often find that iris fail to bloom after they've been divided. The secret is to plant them so that the top of the rhizome is showing. Barely cover them and they should bloom for you.

Hostas prefer moist, well-drained soil, but I've seen them do well in a wide variety of conditions. Most of them are very hardy, tolerating zone 3 temperatures, but check at your local garden centre for the hardiness of plants before you buy.

After a few years hostas show a noticeable decline in vigour. At this point, get out the fork and divide them while they're dormant. They need few nutrients once they get growing and, as the various clumps fill in, they create a magnificent display.

Slugs and snails are always a problem because they love to munch on hosta leaves, which really spoils their appearance. Beneficial nematodes bred to attack slugs may be the most promising hope for the future when it comes to protecting hostas from nightly attacks.

There are hundreds of types of hostas adapted for use in different situations, and they can be categorized in the following groups.

Dwarf. 20 cm (8 inches) or less; slow growing.

Edging. 30 cm (12 inches) or less; vigorous growers.

Groundcover. 45 cm (18 inches) or less; excellent for mass planting.

Background. 60 cm (24 inches) or taller.

Specimen. Usually larger than 60 cm (24 inches), or have unique characteristics.

For details on hostas for specific uses, consult the main listing below.

- 'Big Daddy' (background); 75 cm (2½ feet) high; quilted, deep blue leaves; light lavender flowers in July.
- 'Blessings' (dwarf); 15 cm (6 inches); variegated leaves are yellow with white margins; blue flowers in midseason hold colour well.
- 'Blue Angel' (background); 90 cm (3 feet) by 1.2 m (4 feet); huge, textured, very blue leaves; long-blooming, hyacinth-type white flowers. Very tropical and exotic looking— a great overall plant.
- 'Blue Wedgewood' (edging); 35 cm (14 inches) high; heavy, frosty grey-blue leaves; lavender flowers in July.
- 'Christmas Tree' (specimen); 50 cm (20 inches) by 90 cm (36 inches); distinctive heart-shaped leaves with blue centres and white margins; soft lavender flowers. Prefers shade. The leaves grow all the way up the flower stems, creating the Christmas tree effect.
- 'Fragrant Bouquet' (specimen); 45 to 70 cm (18 to 28 inches) high; leaves have bright, apple-green centres with light yellow margins; fragrant white flowers with a light blue fringe bloom in profusion in July and August. Sun tolerant and the 1998 Hosta of the Year.
- 'Frances Williams' (specimen or background); 80 cm (31 inches); round, puckered, soft blue-green leaves with wide yellow margins; pale lavender flowers. Outstanding appearance.
- 'Gold Edger' (edging); 25 cm (10 inches) high; chartreuse to gold, heart-shaped leaves. Rapid grower that's pest tolerant.
- 'Gold Standard' (groundcover); 60 cm (24 inches); yellow base with green margins at prime, when leaves mature; pale lavender flowers midsummer. Some sun is needed to bring out the best gold colouring. Multiple award-winner.
- 'Great Expectations' (specimen); 40 to 45 cm (16 to 18 inches); puckered, deep blue leaves with bold centre variegations of golden yellow to creamy white. Very exotic.
- 'Halcyon' (groundcover); 50 cm (20 inches) high by 90 cm (3 feet) wide; chalky, thick bright blue leaves; pink-lavender flowers in August. Moderate grower, good for containers.
- 'Honey Bells' (specimen); 65 cm (26 inches) high by 75 cm (30 inches) wide; large light green leaves; pale mauve flowers. Very durable plant.

- *Hosta plantaginea* (specimen); 60 cm (2 feet) high by 90 cm (3 feet) wide; shiny, light green leaves; huge, fragrant, pure white flowers in August. Exceptionally beautiful and one of the original hostas.
- *Hosta sieboldiana* 'Elegans' (background); 1 m (3 feet) by 1 m (3 feet); large, corrugated, frosted powder-blue leaves; white flowers with a hint of lavender in early summer. Eventually grows huge, but needs rich soil and shade to do well.
- 'Invincible' (edging), 25 cm (10 inches) high; glossy, bright green leaves that are wavy and pointed; long-blooming, fragrant

Best bets

Sun-tolerant hostas
Even with sun-tolerant hostas, expect some fading of colour in intense sun and a bit of burn if the weather changes suddenly from cloudy to brilliant hot sun.
- 'Fragrant Bouquet'
- 'Gold Edger'
- *Hosta plantaginea*
- 'Invincible'
- 'On Stage'
- 'Royal Standard'

Slug- and snail-tolerant hostas
Very thick-leafed varieties are usually less appealing to slugs and snails, but when they're hungry expect some damage.
- 'Big Daddy'
- 'Blue Wedgewood'
- 'Halcyon'
- 'Krossa Regal'
- 'Sum and Substance'

Fragrant flowers
These all smell great when you get up close and sniff the flower, but they seldom fill the air with fragrance.
- 'Fragrant Bouquet'
- 'Honey Bells'
- *Hosta plantaginea*
- 'Invincible'
- 'Royal Standard'
- 'So Sweet'

Blue hostas
- 'Blue Angel'
- 'Halcyon'
- *Sieboldiana* 'Elegans'
- 'Love Pat'

Gold hostas
Gold and chartreuse are wonderful accent colours in the garden and gold-coloured hosta leaves spark up shady areas.
- 'Blessings'
- 'Gold Standard'
- 'Piedmont Gold'
- 'Shademaster'
- 'Sun Power'

Overall performance
- 'Christmas Tree'
- 'Frances Williams'
- 'Gold Standard'
- 'Great Expectations'
- 'Invincible'
- 'Minuteman'
- 'Patriot'
- 'Sum and Substance'
- 'Wide Brim'

lavender flowers in August. Sun tolerant and slug resistant.

- 'Krossa Regal' (specimen); 75 to 90 cm (2½ to 3 feet); fabulous, powder blue–green leaves that are very heavy and leathery; lavender flowers in midsummer.
- 'Love Pat' (specimen); 50 cm (20 inches) by 60 cm (24 inches); good blue that holds its colour even in some sun; pale lavender flowers in summer; puckered round foliage. Best in shade with humus-rich soil.
- 'Minuteman' (edging); 30 cm (12 inches) by 60 cm (24 inches); slightly rippled, glossy green leaves edged with a crisp white 2.5 cm (1-inch) margin. Maintains its colour in heat and partial sun.
- 'On Stage' (groundcover or specimen); 45 cm (18 inches) high; two-tone green markings with white centre; soft blue flowers. Needs sun for best variegation.
- 'Patriot' (background); 25 cm (10 inches) by 50 cm (20 inches); white and green foliage pattern; award-winner in 1997.
- 'Piedmont Gold' (background or groundcover); 45 cm (18 inches); gold-chartreuse leaves; white flowers in midsummer. Tolerates some sun and is one of the best groundcovers for shady spots.
- 'Royal Standard' (specimen); 60 cm (2 feet); green-veined leaves; white flowers have a wonderful perfume in August.
- 'Shademaster' (groundcover); 60 cm (24 inches); gold leaves; lavender flowers

 osta companions

It's clear I'm a big fan of hostas, and here are some good hosta companions for striking combinations.

Bergenia varieties. Most varieties work well, especially 'Bressingham Ruby' or 'Bressingham White'. (See "Best bets—foliage perennials" below, for use with other plants.) The evergreen winter foliage and spring flowers add winter excitement and summer contrast to your hostas.

Carex elata 'Bowles Golden'. This gold and green grasslike perennial makes a wonderful border or edge planting around all hostas, particularly the gold or blue varieties.

Foamflower (Tiarella species). Early flowers, green contrasting foliage and bronze-red winter colour make them natural buddies for hostas. (See "Best bets—perennials for a shady spot" above.)

Helleborus species. Flowers and foliage add winter interest to the hosta garden. *H. foetidus* has beautiful summer leaves.

Japanese anemones (Anemone x hybrida). White and pink flowers that bloom through late summer into fall will give your hosta garden a lift at the end of the season.

Japanese painted fern (Athyrium nipponicum 'Pictum'). Spectacular silver-bronze leaves and reddish stems add spice to any hosta collection.

Lungwort (Pulmonaria species). There are many great new varieties with exceptional foliage and early flowers. (See "Best bets—perennials for a shady spot" above.)

midsummer. Great for weed control because it's fast growing.

- 'So Sweet' (edging); compact 35 cm (14 inches) high; white margins on green leaves; fragrant white-lavender flowers in August; good sun tolerance. Award-winner for 1996.
- 'Sum and Substance' (background); 75 cm (30 inches) high; enormous, textured, chartreuse to gold leaves; lavender flowers bloom late into summer. It's sun tolerant and slug resistant.
- 'Sun Power' (background); 75 cm (30 inches); twisting gold leaves colour up early and hold through fall; light, orchid-type flowers midsummer. Fast growing and needs a little sun for best colour.
- 'Wide Brim' (specimen); 55 cm (22 inches); roundish blue-green leaves with very wide, creamy gold margins have gold tints at maturity; pale lavender flowers in summer.

Best bets—long-flowering perennials

Each perennial has its alloted time to make a display in our gardens, but some have extended blooming times. These are precious plants that are invaluable when it comes to creating groupings with spectacular colour combinations. Much perennial breeding has been focused on improving the length of flowering.

Black-eyed susan (*Rudbeckia fulgida* 'Goldsturm'). Zone 4. Single yellow with brown eye. Blooms July to October. Grows to 90 cm (36 inches). Sun. This is the garden workhorse, standing up to all kinds of weather. It can take a beating and still look great, making it a cornerstone of the summer garden. A classic with ornamental grasses.

Bloody cranesbill (*Geranium sanguineum* 'Shepherd's Warning'). Zone 5. Pink flowers bloom June to September. Grows to 30 cm (12 inches). Sun to part shade. A favourite because of its compact habit and continual flowering.

Common sneezeweed (*Helenium autumnale*). Zone 4. Yellow, orange or red blooms July to September. Grows to 1 to 1.2 m (3 to 4 feet). Sun. This North American native is invaluable for late summer and fall colour. It loves water but tolerates heat well. Cut it back in early June to really bulk it up for a spectacular fall display.

Daylily (*Hemerocallis* 'Stella d'Oro'). Zone 5. Yellow flowers bloom June to October. Grows to 45 cm (18 inches). Sun to part shade. After almost two decades in our gardens, I have never seen 'Stella d'Oro' without flowers from summer until frost. It's a neat and tidy daylily that doesn't look messy after it flowers.

Lewisia (*Lewisia cotyledon*). Zone 3. Pink, yellow or orange blooms May to August. Grows to 10 cm (4 inches). Sun to part shade. It's hard to find a longer-blooming and more stunning rockery perennial. Many spectacular new varieties are coming out of Europe, such as the Ashwood Strain. They need exceptional drainage and I suggest you put some pea gravel under their ground-hugging leaves to prevent rot. The payoff is nonstop colour.

Pincushion flower (*Scabiosa columbaria* 'Pink Mist' and 'Butterfly Blue'). Zone 5. Lavender-blue or soft pink blooms from May to October. Grows to 30 cm (12 inches). Sun. Double anemonelike flowers attract bees and butterflies. I saw them overwinter in Washington State and I was surprised to see how compact and tidy they looked even in containers. In zone 6 or above, we could classify them as evergreen. In wet springs, a few applications of fungicide will help prevent mildew, but planting them in well-drained sunny areas with good air flow will help.

Pink-flowered coreopsis (*Coreopsis rosea* 'American Dream'). Zone 4. Pink flowers bloom July to September. Grows to 30 cm

(12 inches). A refreshing, bright pink daisy to cool down those hot summer colours. It was the 1993 Perennial Plant of the Year. Likes sun but will tolerate light shade.

Purple coneflower (*Echinacea purpurea*). Zone 4. Purple or white flowers July to September. Grows to 1 m (40 inches). Sun. A field in full bloom will steal your heart forever with its magical rosy pinks. It makes a great cut flower, dries well, and attracts butterflies and bees. You may have seen it in the vitamin supplements section of your drugstore as an immune system booster.

Tickseed (*Coreopsis verticillata* 'Moonbeam'). Zone 4. Blooms July to September. Grows to 60 cm (24 inches). Sun. Small, soft yellow flowers on airy stalks blend beautifully with dark foliage plants. It gets a little sloppy though, drooping down when it rains.

Tickseed (*Coreopsis verticillata* 'Zagreb'). Zone 4. Bright yellow flowers from June until frost. Grows to 45 cm (18 inches). Sun. A

Best mildew-proof garden phlox

Tall, majestic, summer-blooming *Phlox paniculata* is truly one of the garden highlights. Rain showers, however, can spoil the show by causing powdery mildew to appear all over the leaves. You can control mildew to some extent by using a systemic fungicide, but these products are preventatives, so they must be applied before there is a problem, every seven to ten days during wet weather. At some point, I drew a line and decided enough was enough with battling mildew. Out went the old mildew-prone varieties and in went the mildew-resistant ones. Here are my picks.

- *Phlox* x *arendsii*. Zone 3. 'Anja' has dark pink flowers on 50- to 60-cm (20- to 24-inch) plants. Starts to bloom in June and flowers for most of the summer. 'Hilda' is like 'Anja', but with bicolour white and pink flowers. 'Susanne' has bicolour white and red flowers.
- *P. carolina* 'Miss Lingard'. Zone 5. Pure white, fragrant flowers on shorter, 50-cm (20-inch) stems. One of the longest blooming varieties and wonderful for cut flowers.
- *P. maculata* 'Natascha'. Zone 5. Pink and white flowers on 50-cm (20-inch) stems. Ideal as cut flowers.
- *P. paniculata* 'Andre' has sky blue flowers on 80-cm (32-inch) stems. 'David' has exceptionally large, white flowerheads on 105-cm (42-inch) stems. 'Flamingo' has very fragrant, striking pink flowers with a softer pink eye on 65-cm (26-inch) stems. 'Laura' has purple flowers with a white star centre on 60-cm (24-inch) stems. 'Mini Star' is a dwarf variety, on 30-cm (12-inch) stems and has rose flowers with a darker red eye. 'Miss Ellie' has flamed dark red to dark pink flowers; dark foliage on 70-cm (28-inch) stems sets this one apart. 'Miss Kelly' has magnificent lilac flowers on 70-cm (28-inch) stems. 'Nicky' is the darkest purple of all phlox varieties on 75-cm (30-inch) stems. All are zone 4.

long-lived, drought-tolerant, carefree perennial. It combines well with dark-leafed heucheras and ornamental grasses.

Yarrow (*Achillea* 'Summer Pastels'). Zone 4. Blooms June to September. Grows to 90 cm (36 inches). Sun. I first saw this wonderful blend of pastel shades back in the '80s when it was introduced. I was impressed with it then and am even more impressed now. The colour range harmonizes beautifully with most other plants, and its medium height allows it to be used between low and tall companions.

Best bets—foliage perennials

In the past we've planted perennials mostly for their flower colour and form, but with today's emphasis on year-round colour, foliage has become part of the decision-making process when choosing plants. Leaves that contrast nicely with flowers enhance the plant's appearance and colourful foliage adds value far beyond the flowering period. Attractive leaves are invaluable in container plantings, especially in winter. The most wonderful asset of the new foliage perennials is their use in combination with other plants.

Remember that the foliage of most perennials changes with the seasons, especially when new growth appears and when the dormant period approaches. In regions with cold winters, where the temperature drops below −15° to −20°C (4° to 5°F), most tender evergreens, like lungworts and spurges, will lose their leaves until spring.

Bergenia (*Bergenia* 'Bressingham Ruby'). Zone 3. Bronze-red in fall, winter and spring; green in summer. Grows to 30 cm (12 inches). Sun to part shade. I spotted this in Adrian Bloom's garden in England in late September one year and the rich burgundy

Perennials for dried flowers

Dried flowers have always been a down-home favourite. These varieties all need sunshine and I find well-drained, sandy soils help the seed pods mature more quickly.

Baby's breath (*Gypsophila paniculata*). Zone 4. The single-flowered varieties dry best. Needs a hot, sunny location with superb drainage.

Chinese lantern plant (*Physalis alkekengii*). Zone 5. If you have a place where it can spread in the sunshine, you'll love the bright orange flowers.

Globe thistle (*Echinops ritro*). Zone 4. Globe-shaped, metallic-blue, spiky flowerheads are fantastic when dried.

Maiden grass (*Miscanthus sinensis*) and fountain grass (*Pennisetum alopecuroides*). Zone 4. Two grasses with wonderful flowerheads that look great in any dried arrangement.

Sea holly (*Eryngium*). Zone 5. The amazing flowers have a delicate, lacy, silvery appearance, but are prickly to the touch. *E. alpinum*, with blue-silver tinges, is a great variety.

Tatarian statice (*Goniolimon tataricum*). Zone 4. The silver-white German statice is the most popular, but sea lavender, or statice, (*Limonium latifolium*), with its low, blue panicle, is fast becoming a favourite.

foliage highlighted everything around it. It's a winter jewel when frost is on the fringes. It makes a great groundcover.

Coral bells (*Heuchera* 'Plum Pudding'). Zone 4. Shiny plum-purple foliage. Grows to 30 cm (12 inches). This is my favourite of all the dozens of dark-leafed varieties. In sun or partial shade, the rich plum foliage accents everything around it.

Foamflower (*Tiarella cordifolia*). (See "Best bets—perennials for a shady spot" above.)

Lady's mantle (*Alchemilla mollis*). Zone 2. Chartreuse foliage and chartreuse blooms from June to August. Grows to 30 to 45 cm (12 to 18 inches). Sun to part shade. Raindrops magically bead on the leaves of this valuable accent plant. Use it to complement flowering shrubs and other perennials or as an underplanting around small trees. Pop some bulbs under the leaves for a spring hit of colour.

Lamb's ears (*Stachys byzantina* 'Silver Carpet'). Zone 3. Grows to 38 to 45 cm (15 to 18 inches). Part shade. We've enjoyed the velvety, silver-grey foliage as a groundcover for two decades, in the same location, year-round. It's a great accent or border plant. You can't walk by without stopping to touch the leaves—I guarantee it!

Rodgersia (*Rodgersia podophylla*). Zone 5. Grows to 1.5 m (5 feet). Sun to part shade. Once this perennial is established, the unusual, large, bronze chestnutlike leaves are a focal point. Creamy white summer flowers are a bonus. It does best in dappled shade and looks fabulous near water.

Spurge (*Euphorbia myrsinites*). Zone 5. Spiky blue leaves and chartreuse flower bracts. Blooms May to June. Grows to 30 cm (12 inches). Sun. This drought-tolerant, sprawling plant is a real eyecatcher, drooping over walls or gracefully spreading out as a foreground planting. It looks great with white bulbs, such as spring snowflake or summer hyacinth.

Stinking hellebore (*Helleborus foetidus*). Zone 5. Green foliage in the spring, bronze in winter. Grows to 30 to 45 cm (12 to 18 inches). Part sun to shade. This is without a doubt the best hellebore for foliage. The dark, rich green, deeply cut leaves turn a wonderful bronze in cool winter temperatures. I love the nodding, light green (but stinky) flowers in late winter or early spring. I've used them to great effect underplanted beneath holly trees.

Wood spurge (*Euphorbia amygdaloides* 'Purpurea'). Zone 5. Bronze foliage in fall, winter and spring; green in summer. Grows to 1 m (3 feet). Sun. Especially beautiful in the spring when the bronze foliage is graced with chartreuse flower bracts. It's at its best under small trees or mixed with other colourful shrubs.

Wormwood (*Artemisia stelleriana* 'Silver Brocade'). Zone 5. Grows to 38 cm (15 inches). Sun. Introduced by UBC Botanical Garden, it became an instant success story. Although it loses its lacy, scalloped silver foliage in winter, it is drought tolerant, neat and tidy. Use it to brighten up dull evergreens. It needs good drainage or it can rot in winter.

Best bets—groundcover perennials

Move over traditional shrubby groundcovers, because here come the perennials! Instead of junipers, laurels and mugho pines, over the past 17 years we have experimented in our gardens with small and large pockets of perennials to cover bare spots and accent the plants around them. My favourite groundcovers are ones that stay green all year, change their foliage colour with the seasons and have self-cleaning blooms.

Some, like *Astilbe chinensis,* will die down in late winter, but their performance the rest of the year more than makes up for this one small concern. Our shady areas are lifted dramatically with the silver *Lamium* varieties, and the rosy red blossoms of *Fragaria* 'Lipstick' are a treat all summer. Trial newcomers for at least one year in small pockets in your garden to make sure their performance meets your expectations. These plants will add a whole new dimension to your landscape.

Barrenwort (*Epimedium* × *rubrum***).** Zone 5. Grows to 30 cm (12 inches). Part sun to shade. This elegant plant is quite at home under rhododendrons or by itself in a semi-shaded area. Reddish flowers top heart-shaped, bronze-tinged leaves that look great from spring through to frost.

Bugleweed (*Ajuga reptans* 'Burgundy Glow').** Zone 3. Grows to 20 cm (8 inches). Sun to part shade. *Ajuga* is making a comeback in popularity, especially with new varieties such as 'Burgundy Glow', with its rich pink, white and burgundy foliage and massive blue flowers in spring. It's great for underplanting around small trees (don't forget to plant spring and fall crocuses for added colour).

Creeping jenny (*Lysimachia nummularia***).** Zone 6. Yellow flowers bloom all summer. Grows to 5 cm (2 inches). Sun to part shade. This is a staple in hanging baskets, but break loose and put it in the ground. We've had it fighting weeds for years on an open south-facing bank, but it works even better in shady spots.

Dwarf Chinese astilbe (*Astilbe chinensis* 'Pumila').** Zone 5. Grows to 8 to 30 cm (3 to 12 inches). Sun to part shade. The foliage is compact and attractive and the mauve flower spikes always open in late summer when we

Fragrant perennials

Colour, form and style are all important elements of a perennial garden, but how about fragrance? A garden should appeal to all our senses, and few things are more enjoyable than walking through your garden during the growing season and catching a drift of some delightful perfume. Always try to strategically locate fragrant perennials where the prevailing breezes will carry the fragrance toward a pathway, an open window, a verandah or outside patio.

Carnations, pinks (*Dianthus spp.***).** Most varieties, single or double, are fragrant. These sun-lovers will flower May through July.

English lavender (*Lavandula angustifolia***).** Walk anywhere near lavender in bloom or simply brush by the foliage to enjoy that refreshing lavender perfume.

Hostas. (See the hostas list earlier in chapter.)

Lily-of-the-valley (*Convallaria majalis***).** This shade-lover produces tiny stems of fragrant, bell-shaped flowers in April and May.

Summer phlox (*Phlox paniculata***).** Elegant plants with summer perfume. (See "Best mildew-proof garden phlox" above, for more details.)

Sweet violets (*Viola odorata***).** Tiny blossoms in March. They love dappled shade.

Wallflower (*Erysimum***).** 'Bowle's Mauve' is not only evergreen in mild climates, but it will bloom in late winter, well into spring and again in the fall. Needs a protected spot even in zone 7.

most need colour. It's hardy, beautiful, and doesn't mind having wet feet.

Himalayan knotweed (*Polygonum affine*). Zone 4. Rosy pink spikes bloom summer to fall. Grows to 45 cm (18 inches). Sun. I first saw this plant used in place of a lawn in Holland—no mowing, and a changing landscape of colour over the course of the year. In the fall, the foliage takes on an attractive red-bronze colour.

Japanese spurge (*Pachysandra terminalis*). Zone 5. Grows to 30 cm (12 inches). Part sun to shade. This is the very best weed-choking groundcover for shade and semishade areas. Rich, evergreen foliage is topped with massive white flowers in spring.

Strawberry plant (*Fragaria* 'Lipstick'). Zone 6. Grows to 30 cm (12 inches). Sun. This looks great draped over walls. It stays evergreen in milder climates, blooms all summer under stressful conditions, and the medium-sized strawberries taste good—a real bonus.

Two-row stonecrop (*Sedum spurium* 'Dragon's Blood'). Zone 5. Red foliage with red blooms in summer. Grows to 15 cm (6 inches). Sun. A drought-tolerant, no-care, hardy plant that's evergreen (or, in this case, ever-red). 'Dragon's Blood' has been used in place of a lawn. Once established it's an attractive, fast-growing groundcover that doesn't require much maintenance.

Woolly thyme (*Thymus pseudolanuginosus*). Zone 4. Seldom flowers. Grows to 2.5 cm (1 inch). Sun. This thyme is the workhorse filler for patio blocks, stone pathways and rock walls. The fuzzy, grey-green foliage is a great accent, you can walk on it and it smells great. I know one fellow who went wild and used it instead of a lawn. He hasn't mowed since.

Yellow archangel (*Lamium galeobdolon* 'Herman's Pride'). Zone 4. Yellow with silver-green foliage. Grows 8 to 30 cm (3 to 12 inches). This distinctive groundcover works well in part sun to shade and has unusual evergreen foliage like the indoor aluminum plant. It stays low, spreads quickly and displays beautiful yellow flowers in spring.

Ornamental Grasses and Grasslike Plants

Grasses and grasslike plants, such as sedges and rushes, are a superb component of the landscape, adding unusual form and texture to the old garden standards. Sometimes our gardens become a little static, but with grasses the slightest breeze creates life and movement. And many grasses, like the tall-growing *Miscanthus sinensis,* add the dimension of sound—that wonderful rustling when the wind blows gently. Plant them near a patio or deck, where you can enjoy the soothing rustling while relaxing in the evening. Once you've seen fall sunshine highlighting the silvery plumes of grasses, you'll be hooked on these fascinating plants.

Grasses can look quite at home in almost any landscape. Adrian Bloom took a liking to grasses and used them in mass plantings throughout his conifer garden, "Foggy Bottom" in England, with the most delightful results. At the Royal Horticultural Society Garden in Wisley, England, grasses add flair and sophistication as focal points in containers and annual beds.

Even in winter, many grasses have a lot to offer until heavy snows eventually take their toll. A long narrow bed of grasses grows against an old split-rail fence in front of our home, and there isn't a time of year I don't enjoy it. In winter, it's especially attractive with its seed heads and straw-coloured stems.

You can find an appropriate grass for almost any use in your garden. Tall grasses can create screens to add a touch of intrigue in your garden, form a neat backdrop for other plants, and even hide those pesky neighbours. Low, fast-spreading ornamental grasses, like mosquito grass (*Bouteloua gracilis*), make interesting groundcovers and are a nice change from some of the traditional plants.

Some sedges, like leatherleaf sedge (*Carex buchananii*), tolerate excessive heat and drought, while others, like variegated sweet flag (*Acorus calamus* 'Variegatus') enjoy wet feet. Grasses such as variegated Japanese sedge (*Carex morrowii* 'Aureo-variegata'), with its silver variegated leaves, brighten a shady spot and combine well with other shade perennials. Many grasses fill niches where other plants have difficulty.

But before you run out and stock up on these versatile plants, there are a few things to watch for. First, determine which grasses are hardy for your area. Some of the most beautiful varieties, like *Pennisetum setaceum* 'Rubrum' with its fabulous red-bronze foliage and long, elegant seed heads, are zone 9–10 plants.

Running grasses can be invasive and need physical barriers to keep them in check. Some of the vigorous grasses may overpower surrounding plantings. *Erianthus ravennae,* while not a running grass, grows very large and needs a large space with plenty of elbow room.

Many grasses are evergreen in milder climates (zone 6 and above), which is another bonus for winter enjoyment. Keep in mind, though, that exceptionally cold winters will knock them down. During most winters, however, evergreen grasses, such as *Hakonechloa macra* 'Aureola', with its bamboolike variegated gold foliage, make wonderful winter accents for dark-leafed perennials like bergenia and deeply coloured winter-flowering heathers.

Grasses with interesting flower plumes make great additions to flower arrangements. The best time to pick them is when the flowers have just emerged from the foliage. If you

𝒯ips for growing grasses

Grasses and grasslike plants are some of the most resilient plants once established, but for best results prepare the soil as you would for any perennial and make sure they have good drainage—unless, of course, they are water lovers.

Grasses can be planted at any time of the year if they are hardy for your area, and well established in containers, although spring and early summer are best to get them ready for their first winter.

Grasses are nitrogen lovers, and once established, slow-release lawn fertilizer is ideal.

Cut back any grass when it begins to spoil the appearance of your landscape, but the best time is in spring, just before the new growth begins. Some fast-growing and sloppy grasses can be cut back during the growing season to produce fresh new growth and a new start. Prune about 2.5 to 5 cm (1 to 2 inches) above the crown, although sedges and many evergreen grasses should not be pruned quite so hard.

To keep your grasses looking good during hot spells, water deeply and thoroughly; if you don't, they will tend to move to fall colouring more quickly. If grasses simply get too big and sloppy, don't be afraid to dig them up, divide and replant. It will give them a whole new lease on life.

wait too long, they will shatter soon after picking. Cut grasses during the warmest part of the day, after the morning dew has dried and before the cool dampness of evening. Grasses for use in fresh arrangements should be put in water immediately. If you're using them in a dried arrangement, hang them upside down in a cool, dry, dark place.

Best bets—screens or tall specimens

Giant reed grass (*Arundo donax*). Zone 6. This amazing plant grows 3 to 7.5 m (10 to 25 feet) and has blue-green foliage up to 60 cm (2 feet) long, with 8-cm-wide (3-inch), hollow, canelike stems. Large fluffy panicles appear in September and last into winter. It's well suited to water edge plantings, but you do need some space. Flowers are beautiful in arrangements.

Japanese silver grass (*Miscanthus sinensis* varieties). Zone 5. Sun. These are among the most beautiful of today's clumping grasses; they grow from 1.2 to 2.1 m (4 to 7 feet), depending on variety. It puts on a good garden display by late May and bears showy, whisklike plumes from September to October. 'Purpurascens' has brilliant, red-orange fall foliage. Nice accent for winter or as a border.

Mauritania vine reed (*Ampelodesmos mauritanica*). Zone 6. Sun. This fast-growing clumping grass grows to 1.2 to 1.8 m (4 to 6 feet) with feathery plumes from July to December. It needs room. It makes a great cut flower.

Ravenna grass (*Erianthus ravennae*). Zone 5. Sun. This clumping deciduous grass is very fast growing with foliage reaching 1.5 m (5 feet). It has tall, narrow, silvery flowers with purple tones, reaching up to 3.6 m (12 feet) from August to December.

Best bets—groundcovers

Blue fescue (*Festuca cinerea* 'Elijah Blue'). Zone 4. Grows to 25 to 40 cm (10 to 16 inches). Sun. This is one of the bluest forms of this clumping evergreen fescue. Thin, spiky, straw-coloured flowers appear in late summer and are usually trimmed to maintain an attractive look. They do best in moist, but well-drained, soils and the whole plant should be pruned back to 10 to 13 cm (4 to 5 inches) in spring for the very best appearance.

Dwarf fountain grass (*Pennisetum alopecuroides* 'Hameln'). Zone 6. Grows to 40 to 60 cm (16 to 24 inches). Sun. Neat, compact grass with slender leaves and whitish flowers in late summer. It does particularly well in moist climates with cool winters.

Feather grass (*Stipa capillata*). Zone 6. Grows to 50 to 100 cm (20 to 40 inches). Sun. The outstanding feature of this semi-deciduous clumping grass is its silky masses of tiny silvery flowers in July and August. The grass is evergreen in mild climates (zone 6 and above). It's a wonderful accent plant and ideal in fresh or dried arrangements.

Mosquito grass (*Bouteloua gracilis*). Zone 4. Sun. Grows to 30 to 50 cm (12 to 20 inches). Tiny comblike flowers, resembling mosquito larvae, are white when young and then turn bronze-purple. You can mow it and it tolerates drought. Flowers can be used in fresh or dried arrangements.

Sideoats grass (*Bouteloua curtipendula*). Zone 4. Grows 50 to 80 cm (20 to 32 inches). Clumping, drought-tolerant grass. Flowers appear on one side of the stem in June. Great on slopes as erosion control. Flowers can be used in dried or fresh arrangements.

Best bets—bog or shallow water specimens

Banded zebra rush (*Schoenoplectus tabernaemontanus* 'Zebrinus'). Zone 6. Grows to 60 to 120 cm (2 to 4 feet) high. Attractive hollow stems have yellowish bands. At its best in spring; it fades by late summer. The foliage turns yellow in fall, then brown in winter.

Blue-green rush (*Juncus inflexus*). Zone 4. Sun or partial shade. This evergreen clumper loves boggy areas and grows 60 cm (24 inches) tall with deeply striated, almost glaucous stems. Brown flowers are borne on the tips of the stems in early summer and persist until fall. Great for pots and water tubs.

Cyperus sedge (*Carex pseudocyperus*). Zone 5. As its name suggests, this deciduous sedge resembles papyrus. It has yellowish green foliage and 60- to 90-cm (2- to 3-foot) spikes with light brown flowers.

Golden-edged prairie cord grass (*Spartina pectinata* 'Aureomarginata'). Zone 4. Glossy green foliage with a gold margin grows 1 to 1.8 m (3 to 6 feet) tall. In late July, one-sided brownish spikes are borne about 1 m (3 feet) above the leaves, and in the fall, the foliage becomes golden yellow.

Palm sedge (*Carex muskingumensis*). Zone 4. This semideciduous sedge has arching 60- to 90-cm (2- to 3-foot) stems with palm-like, 20-cm (8-inch) leaves. It's an excellent groundcover for moist banks. It also looks great as a water feature.

Spiral rush (*Juncus effusus* 'Unicorn'). Zone 7. Grows to 45 to 60 cm (18 to 24 inches). This is a real conversation piece in wet tubs or shallow ponds. It has the richest colour in partial shade, and it must never become bone dry. Its spiraling, corkscrew stems are sought after by floral designers.

Best bets—shade-lovers

Silver variegated Japanese sedge (*Carex morrowii* 'Variegata'). Zone 6. Grows to 45 to 60 cm (18 to 24 inches). This creamy yellow variegated plant is evergreen and looks fabulous planted with other perennials under trees. Flowers are not significant.

Snowy woodrush (*Luzula nivea*). Zone 4. Grows to 20 to 30 cm (8 to 12 inches). Grey-green evergreen foliage is covered in soft downy hairs. Its showy white flowers are borne about 30 cm (12 inches) above the foliage. The fluffy flowerheads last into late fall.

Strawberries and cream ribbon grass (*Phalaris arundinacea* 'Feesey's Form'). Zone 4. Variegated white and green foliage with a pink blush makes a great summer accent. Soft white flowers carried 30 cm (12 inches) above the foliage are very showy in winter.

Tufted hair grass (*Deschampsia caespitosa*). Zone 4. This handsome evergreen with its dark green foliage grows up to 1 m (3 feet) and produces silken, airy flowers that almost smother the foliage. They mature to a golden bronze and are showy well into summer. One of the earliest grasses to bloom, they are ideal companions for hostas.

Best bets—compact garden specimens

Black-flowering pennisetum (*Pennisetum alopecuroides* 'Moudry'). Zone 6. Grows to 60 to 80 cm (24 to 32 inches). A striking plant with lush green foliage that sports a large black flower September through fall.

Blue oat grass (*Helictotrichon sempervirens*). Zone 4. Grows to 60 to 75 cm (24 to 30 inches). Bright blue foliage with showy flowers in June. This plant is a must. It's evergreen in zone 6 and is a great accent. Needs well-drained fertile soil and full sun.

Bulbous oat grass (*Arrhenatherum elatius* 'Variegatum'). Zone 4. This white striped grass grows only 15 to 30 cm (6 to 12 inches), with small floral spikes appearing in May and June. They look great in rockeries and perennial borders and are particularly striking when combined with dark-leafed heucheras and silver artemisias.

Golden hakone grass (*Hakonechloa macra* 'Aureola'). Zone 5. This gold-striped evergreen has arching foliage that turns to intense pink tones in fall. It is slow growing to 30 to 60 cm (12 to 24 inches) and is well suited to containers, edgings and rock gardens. It's a keeper. Needs partial shade for best performance.

Japanese blood grass (*Imperata cylindrica* 'Rubra'). Zone 6. The bright green leaves of spring are tipped with red and grow redder as they mature. It doesn't produce flowers. This grass looks good anywhere in the landscape but is particularly nice in containers.

Leatherleaf sedge (*Carex buchananii*). Zone 7. An evergreen sedge with narrow, leathery leaves and coppery brown stems that look good all year. This grass only grows to 45 to 60 cm (18 to 24 inches) tall and with its narrow form provides a great accent to silver, bronze and dark-leafed perennials. It looks great in containers as a focal point.

Miniature cattail (*Typha minima*). Zone 4. Grows to 10 to 15 cm (4 to 6 inches). Very stiff green foliage producing cute miniature cattails in late summer. Prefers moist sites but is very versatile. It is a nice addition to pots and water tubs and works as a groundcover between stepping stones.

Mondo grass (*Ophiopogon planiscapus* 'Nigrescens'). Zone 6. Clumps of black, straplike leaves are a wonderful contrast with silver foliage or white-flowering plants. They need very well-drained soil and prefer sun. They are slow growing, but once established they make a nice display when accented with silver. Good small container plant.

New Zealand hair sedge (*Carex comans* 'Frosty Curls'). Zone 7. Light green foliage. Loves sun and partial shade and spills wonderfully over rocks, walls and containers. It grows up to 1 m (3 feet), but will get larger under the right conditions.

Bronze New Zealand hair sedge (*Carex comans* 'Bronze'). Zone 7. Fine, brownish white foliage only grows to 20 cm (8 inches). It's ideal for small containers, rockeries and heather beds.

Oriental fountain grass (*Pennisetum orientale*). Zone 7. Grows to 30 to 50 cm (12 to 20 inches). This small grass is ideal in borders, with its glossy blue-green leaves and violet-pink flowers. It is a wonderful winter plant.

***Pennisetum alopecuroides* 'Little Bunny'.** Zone 7. This is the most compact of the fountain grasses—only 10 to 15 cm (4 to 6 inches) high, with white flower spikes that turn tan in late summer. Ideal for small space plantings and containers.

Best bets—mid-sized garden specimens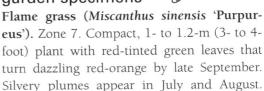

Flame grass (*Miscanthus sinensis* 'Purpureus'). Zone 7. Compact, 1- to 1.2-m (3- to 4-foot) plant with red-tinted green leaves that turn dazzling red-orange by late September. Silvery plumes appear in July and August. Remains attractive into winter.

'Heavy Metal' switch grass (*Panicum virgatum* 'Heavy Metal'). Zone 5. Grows to about 1 m (3 to 3½ feet). Blue-tinged leaves turn bright yellow in the fall. Bears distinctive panicles of light brown seed heads.

Japanese silver grass (*Miscanthus sinensis* 'Morning Light'). Zone 5. Grows to 1.5 m (5 feet). White leaf margins give it a silvery appearance. Flowers bloom reddish bronze in October and turn creamy coloured.

Japanese silver grass (*Miscanthus sinensis* 'Variegatus'). Zone 6. Grows to 1.2 to 1.8 m (4 to 6 feet). Wonderful creamy white variegated foliage really lifts slightly shaded spots, although the plant needs some sun. Leaves turn a light tan in fall.

Japanese silver grass (*Miscanthus sinensis* 'Yaku Jima'). Zone 5. This fine-textured grass is suitable for small gardens as it grows to only 1 to 1.2 m (3 to 4 feet). Leaves turn reddish brown in the fall. Flowers are reddish at emergence, becoming silver in the fall. They stay within the foliage, but are so full they almost obscure the leaves.

Porcupine grass (*Miscanthus sinensis* **'Strictus'**). Zone 4. Grows to 1.2 to 1.8 m (4 to 6 feet) high. This grass has stiff, upright green leaves with horizontal yellow bands that extend to the leaf tips. A good accent plant.

Red switch grass (*Panicum virgatum* **'Haense Herms'**). Zone 5. Compact habit, about 1 m (3 to 3½ feet). Leaves turn red-orange in the fall, eventually becoming grey-brown. Seed heads create a smoky effect.

7 Roses

The rose has been cultivated for thousands of years and is the most popular garden plant in the world, bar none. In selecting a rose for your garden you have around 20,000 types to choose from, as well as several hundred new ones that appear on the market each year. There are endless combinations of size, colour, leaf and petal shape to choose from, as well as hardiness and fragrance to take into consideration. It just goes to show that "a rose is a rose is a rose" isn't necessarily so. I don't think this wealth of choice has to be intimidating; think of it as an acknowledgement of how important the rose is to the gardener.

With the resurgence of interest in old roses from the past, an explosion of modern varieties appearing every year, the reintroduction of species roses, and the recent development of David Austin English roses, there are a lot to choose from. In the end, most of us want a colourful, fragrant, long-blooming rose that's disease resistant and pest tolerant. Roses can be divided into three simple and broad classifications: species roses, old garden roses and modern roses. From there you can get into many different subgroups (more on that later).

Rose flowers are described as single, semidouble or double. Single varieties produce five petals on each flower, while semidoubles have a few more than five, but not so many that the centre is obscured. Double-flowering varieties have many petals, giving the flower a much fuller appearance.

The following pages detail some of the many different types of roses, including old-fashioned varieties, such as the damasks and albas, the modern hybrid teas and floribundas, David Austin roses and others. At the end of the section are some tips on rose care, such as how to winterize and prune your roses.

Species Roses

Species roses are native roses found growing in the wild, which have since been put under cultivation. There are well over 150 species of roses currently under cultivation, but it's sometimes difficult to determine which are true natives, and which have been cross-bred by rose breeders to improve the original characteristics of the plant. The distinct characteristics of species roses are their mostly single-flower, disease-free nature and colourful fall seed pods, known as rose hips. Most are very cold hardy and can be left to sprawl up trees or over old fences, buildings and even hedges.

These native roses and selections from them have been treasured for centuries because of their perfume, disease tolerance and fall seed hips. They range in size anywhere from 60 cm (24 inches) to over 15 m (50 feet). The flowers are either on single stems or in clusters and range from yellows and whites to pinks and dark reds. Each flower usually has five petals.

- *Rosa banksiae* (Lady Banks' rose). Zone 7. Masses of tiny, double, very fragrant white flowers appear in summer only, but the leafy, pale green stems reach up to 6.7 m (22 feet), making it ideal for training up trees, old buildings, arbours or across a trellis.
- *Rosa glauca* (*R. rubrifolia*). Zone 4. Violet foliage on young stems fades to a purplish tone in midsummer if planted in a sunny location. Abundant flowers have single, light-to-dark-pink petals centred with beautiful yellow stamens. Produces small red fruits in fall. Grows to 3 m (10 feet) and provides a nice backdrop. Watch for rust problems.
- *Rosa sericea* f. *pteracantha*. Zone 6. Fernlike foliage, spectacular, translucent red thorns on new growth and unusual, four-petaled, single white blossoms with raised gold stamens make this rose unique. Grows to 2.4 m (8 feet) in height. It needs a sunny spot, and although it blooms only in summer, the thorns are a conversation piece, particularly when backlit by the sun.

Old Garden Roses

Old garden roses are very much in vogue today because of their old-world appearance, strong fragrance and disease tolerance. This grouping refers to roses originally bred prior to 1867. That's the year chosen as the demarcation point between old roses and modern hybrid tea roses—that year 'La France', a silvery lilac-pink, was introduced. But rose history goes back much further than 1867. The oldest of the old garden roses are perhaps the gallicas, grown by the early Greeks and Romans. In the 17th century Europeans took to roses with a passion and began breeding them on a fairly large scale.

Old garden roses include the albas, bourbons, centifolias, China roses, damasks, gallicas, hybrid perpetuals, moss, noisettes, Portlands, sempervivens and teas. Following are a selection of some of these rose groups that I feel make significant contributions to our modern gardens. *Note:* NR = non-repeat-flowering; R = repeat-flowering.

Personal picks—gallicas

The gallicas are mostly short, bushy shrubs that are well suited to the home garden. Most bloom only once a season and are noted for their magnificent blends of colour, particularly the deep shades of crimson, purple and mauve. They are hardy to zone 5 without winter protection and seem to thrive in a variety of garden settings. They can be a bit invasive with suckering.

- 'Charles de Mills'. Vigorous, with large, scented, wine-red, closely packed petals in a quartered formation that looks incredibly full. (NR)
- 'Duchesse de Montebello'. Heavily scented, fully double, bluish pink flowers on long, almost thornless, arching canes. (NR)
- 'Ipsilanté'. Zone 4. Very high disease tolerance, light pink flowers with a delightful quartered appearance and terrific fragrance. Shrub grows to 1.5 m (5 feet). (NR)
- 'Officinalis'. An ancient variety often called the "apothecary's rose." It has large, semi-double, bright crimson flowers that reveal beautiful gold stamens when they open. It makes a spectacular show in full bloom.

The fragrant petals have been a source of potpourri for centuries. A sprawling shrub to 1.5 m (5 feet), it bears red hips in autumn.

Personal picks—damasks

Many people suggest that the damask roses were brought to Europe by the Crusaders. This rose group is noted for deeply cut foliage, elegant growth patterns and fragrant flowers held in open, airy clusters. Their deep, long-lasting fragrance is often referred to as "expensive." Most are hardy, easy to grow, and have excellent disease tolerance. Remove some old wood each year to stimulate new stems.

- 'Celsiana'. Zone 4. Long, arching branches allow you to use this shrub as a climber if you wish. Smaller clusters of pink, semi-double flowers with exquisite golden anthers. Blooms heavily to fill the garden with a delightful perfume. Grows to 1.5 m (5 feet). (NR)
- 'Mme. Hardy'. Zone 4. An outstanding shrub rose to 1.5 m (5 feet), with a hint of lemon fragrance and a green button in the centre of its pure white, double blossoms. (NR)
- 'Quatre Saisons', or 'Autumn Damask'. Zone 4. Loose, double pink flowers produce a beautiful and intense perfume. Very reliable under diverse weather conditions. Repeat blooms in autumn—the oldest damask to repeat flower. Grows to a 1.5-m (5-foot) shrub. (R)

Personal picks—albas

Alba roses (*Rosa* × *alba*) date back to the Middle Ages. They came to fame during the 15th century when an alba rose represented the House of York during the War of the Roses, while the House of Lancaster stood behind a red rose (*R. gallica* 'Officinalis'). The alba colour range is limited to soft pink, blush and white, but they add elegance to any garden. They are very disease tolerant and survive under adverse conditions, including partial shade. They require only minimal pruning; take out the old wood after flowering to encourage new growth.

- 'Great Maiden's Blush', or 'Cuisse de Nymphe'. Zone 4. A large shrub—to 2.1 m (7 feet)—that's often used as a hedge. Abundant, blush-pink, sweetly scented blooms. Attractive blue-grey foliage on long arching branches that are often weighted down during flowering. (NR)
- 'Maxima'. Zone 4. One of the most popular alba roses because of its ability to survive and perform under the most difficult conditions. Bluish pink flowers fade to creamy white when open and are very fragrant. Shrub to 2.1 m (7 feet) bears orange-red hips in fall. (NR)
- 'Queen of Denmark', or 'Konigin von Dänemark'. Zone 4. Reputed to be the finest of all old roses, with its quartered, soft pink, double flowers and strong scent. Grey-green leaves contrast beautifully with the flower colour. Long blooming period. A shrub that grows to 1.5 m (5 feet). (NR)

Personal picks—centifolias

Centifolias (*Rosa* × *centifolia*) are also complex hybrids originating in Europe and include some of the most opulent of the old-fashioned roses. They bloom only once a year, but are very hardy and have reasonable disease tolerance. Their downside is a more open and sparse habit and massive thorns, but the intense perfume and rich colour of the huge blossoms make them well suited to a cottage-type garden.

A frequent subject of the Dutch master painters, the loose, open blossoms have unflatteringly been dubbed "cabbage roses." The size and weight of the blossoms causes many varieties to hang their heads or nod a bit, giving them an unusual, but pleasing look in the

garden. After flowering, cut back growth by one-half, removing all weak, twiggy wood.
- 'The Bishop'. Zone 4. Highly perfumed, double, magenta-cerise blossoms fading to an attractive bluish grey. Needs full sun. Suitable for containers. Shrub to 1.5 m (5 feet). (NR)
- 'Pomponia', or 'Rose de Meaux'. Zone 4. Highly scented, clear pink, cup-type rose is the original, loose-form cabbage rose. A shrub that grows to 1 m (3 feet). (NR)

Personal picks—moss roses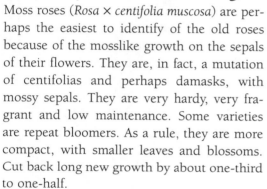

Moss roses (*Rosa* × *centifolia muscosa*) are perhaps the easiest to identify of the old roses because of the mosslike growth on the sepals of their flowers. They are, in fact, a mutation of centifolias and perhaps damasks, with mossy sepals. They are very hardy, very fragrant and low maintenance. Some varieties are repeat bloomers. As a rule, they are more compact, with smaller leaves and blossoms. Cut back long new growth by about one-third to one-half.
- 'Chapeau de Napoléon'. Zone 4. Unusually beautiful, mossy buds resembling an 18th-century cocked hat—hence the name. Very large, pure pink, richly fragrant flowers. Shrub grows to 1.2 m (4 feet). (NR)
- 'Mousseline'. Zone 4. Large, semidouble blush-pink blossoms. Has a sweet mossy scent. Grows to 1.2 m (4 feet). (R—for up to 6 months)
- *Rosa* × *centifolia muscosa* (common moss rose). Zone 4. Double, clear pink flowers have a deep, rich perfume. Grows to 1.5 m (5 feet). (NR)

More old garden roses

Those above are the main groups of old garden roses, but the roses listed below have some important characteristics. They have foliage more like modern roses, and many have the ability to bloom repeatedly, with flowers that resemble those of the old varieties in form and fragrance. The aim of this book is to create awareness of these many types, in hopes that you will discover their exquisite beauty for your garden.

Bourbon roses. These roses are thought to have originated from a cross between an old China rose and the damask 'Quatre Saisons'. This group produces long climberlike canes. Its glossy foliage has reddish hues, and the smaller flowers appear in clusters of three to seven. The colour range of bourbons is from soft to dark pink, some with interesting stripes. They grow anywhere from 1 to 3 m (3 to 10 feet) and are hardy in zone 5.

Some of the best repeat-bloomers in this class are the pure white, camellialike flowers of fragrant 'Boule de Neige'; the large, double-flowered blossoms, soft pink striped with lilac and purple, of 'Honorine de Brabant'; and the fully double, quartered, sweetly scented blush-white 'Souvenir de la Malmaison'.

China roses. Zone 6 to 9. Hybrids of *Rosa chinensis* grown in the gardens of China as far back as the 10th century, the main feature of this class is its ability to flower early and several times a season. These roses are generally small, to 1 m (3 feet), with light green, red-tinged foliage and a dense twiggy habit. There is a wide range of colours from whites and pinks to scarlet. Prune back up to 50 percent in late winter.

One of the finest China roses is 'Old Blush', a small 60-cm (2-foot) bush. Its very free-flowering, semidouble blossoms in shades of rose-pink to soft pink have the fragrance of sweet peas.

Hybrid perpetuals. Zone 6. This group is the closest to modern roses, being a hybrid blend of bourbons and other groups. As the name implies, they are recurrent, but their second flowering is much less prolific. There are a variety of flower forms and a range of colours from pink through crimson. They vary in size from low bushes to climbers. Prune back after flowering each year.

'Frau Karl Druschki' is a wonderful climber with little perfume but a great profusion of white flowers and repeat bloomings. A zone 5, it is one of my favourites.

Noisette roses. Zone 6. These roses are an American gardener's cross between a musk and a China rose hybrid. Very fragrant, medium-sized flowers of white, pale pink and creamy yellow are produced in clusters. Noisettes make great climbers. Old canes should be pruned to allow growth of new wood for the best flowering the following year. Cut back new wood by one-quarter.

'Blush Noisette', one of the earliest noisettes, has blush-pink, mauve and deep pink semidouble flowers that bloom almost continuously. It's a nice climber, growing to 5 m (16 feet), and it tolerates partial shade. 'Mme. Alfred Carrière' produces very fragrant, milky white double blooms continuously. Growing up to 6 m (20 feet), it's wonderful growing along fences, up walls or into trees.

Portland roses. Popular in the early 19th century, these roses gained attention because of their exceptionally perfumed, very double, repeat-blooming flowers. The blossoms, on characteristically short stems, are in shades of deep red to rich pink. For the most part, they are a compact 1 to 1.2 m (3 to 4 feet). All long growth should be cut back by at least one-third in winter. The fully double, quartered pink rosette of 'Madame Knorr', with its wonderful sweet perfume, is one of the better known.

Scotch, or burnet, roses (*Rosa pimpinellifolia*). This is a very hardy species (zone 4) that tolerates poor rocky soils and ocean spray. Plants have dark ferny foliage, thorns to avoid, and they sucker freely. As a rule, hybrids only grow 1 m (39 inches) in height and their flowers, in colours of white, creamy yellow, pink and purple, open like saucers and produce a rather delightful musky scent. They have dark red, almost black hips. Prune only to shape and tidy up. There are various forms with double flowers.

Tea roses. Zone 7. Tea roses are important because they were crossed with hybrid perpetuals to produce our modern hybrid teas. The old tea roses originated in China, probably from a cross between *R. gigantea* and *R. chinensis*. Tea roses are so named because they arrived in England on ships carrying tea imported from East India. These roses are known for their beautiful pointed bud forms, which are held on rather weak stems. They are more tender and come in a wide range of colours. Generally, they grow to 1 m (3 feet) and are considered shrubs.

Old-fashioned rose problems

My only warning about choosing one of the old-fashioned varieties is that some are susceptible to disease, while others can look a bit tatty when not in bloom. With careful selection, though, these problems can be avoided and the payoff is a rose that beats the pants off some of the newer varieties.

To minimize disease problems, plant the most disease-resistant varieties in open sunny areas that have good air circulation. Keep weeds away, and pick off any infected leaves as they appear. Sometimes cutting a rose back and stripping off the leaves is the best way to solve the problem organically, even if it's in the middle of summer. There are organic insecticides and fungicides to control insects and disease, but you must keep a sharp eye out for problems, and use the organic sprays frequently.

'Duchesse de Brabant', a sweet, tea-scented, medium to large bush with cupped pearl-pink blossoms, is one of the best known. It's often used as a hedging rose, and it flowers continuously spring through winter. Prune back after the last hard frost in spring.

Modern Roses

These are the roses I grew up with, and based on my experience they're still the most widely planted and grown varieties across North America and around the world. They can be loosely grouped into hybrid teas, floribundas, grandifloras, miniatures, shrub roses and groundcovers. There are, of course, crossovers in some of these categories, but I'll try to keep things as clear as possible to help you choose the right rose for your situation. (Nothing's ever simple—especially when there are more than 20,000 varieties involved.)

Personal picks—hybrid teas

Exquisite blossoms on long single stems are the trademark of this group, making them ideal for cutting. They bloom in waves from late spring through fall, but need winter protection below zone 7. Most varieties grow up to 1.8 m (6 feet). There is a wide variation in disease resistance, but the following hybrid teas have worked for me.

- 'Broadway'. AARS (All-American Rose Selection) winner in 1986. Exquisite, gold-shaded, pink blossoms, slight fragrance and good disease resistance.
- 'Chicago Peace'. Unusual pink with yellow base on its 13-cm (5-inch) blossoms. Slight perfume, glossy foliage and pretty good disease resistance.
- 'Double Delight'. AARS winner in 1977, with bicoloured, red and cream, 13-cm (5-inch) blossoms. Prolific bloomer and good disease resistance.
- 'First Prize'. Still one of the better deep pinks, with a silver reverse. AARS winner in 1970, fragrant, good disease resistance.
- 'Ingrid Bergman'. One of the best-formed velvety reds, with 13-cm (5-inch), mildly perfumed, long-flowering blossoms. Good disease resistance.
- 'Mister Lincoln'. One of the best reds, with huge, fragrant, 15-cm (6-inch) blossoms. Somewhat of an open-cupped bloomer, but very productive and disease resistant.
- 'Oregold'. AARS winner in 1975, and one of the best deep yellows, with 13-cm (5-inch), fragrant blossoms and good disease resistance.
- 'Pascali'. AARS winner in 1969, and one of the best creamy whites. Has a slight perfume and well-formed, small blossoms. Excellent disease resistance.
- 'Peace'. Introduced in 1945, an AARS winner in 1946, and still one of the world's favourite roses. A soft yellow with pink edges, slight perfume and good disease resistance.
- 'Perfecta', or 'Kordes Perfecta'. Creamy white with deep pink edging. Fragrant, continuous-flowering, good glossy foliage.

Personal picks—floribundas

Floribundas, or cluster-flowered roses, were first developed in 1924 by a Danish breeder who crossed a hybrid tea and a polyantha. These are my personal favourites, with the flower form and foliage of the hybrid teas and the lower growing habit, hardiness and continuous flowering of the polyantha roses. When someone wants a recommendation for a rose bed, my first choice is always a floribunda, because this lower-growing plant with clustered blooms will flower almost continuously from June through October. Floribundas are great for cutting and many have a fine perfume.

- 'Apricot Nectar'. The apricot-pink blossoms with gold at the base are up to 13 cm (5 inches) across and have a beautiful,

heavy, fruity perfume. The dark green foliage is quite disease resistant. A 1966 AARS winner.
- 'Europeana'. This 1968 AARS winner has wonderful, dark foliage and deep crimson, 8-cm (3-inch) blossoms that cover the stems almost down to the ground and just don't quit. Only a slight perfume.
- 'Iceberg'. Its 8-cm (3-inch), pure white, double blossoms and glossy, light green foliage make a wonderful contrast to the deep crimson blossoms and dark foliage of 'Europeana' when planted together. Not a bad perfume and it's quite disease resistant.
- 'La Sevillana'. Lightly fragrant, with brilliant, orange-scarlet, semidouble flowers that persist until frost. Grows to 1.2 m (4 feet) and can be kept at that height with moderate pruning. Good disease resistance.
- 'Pleasure'. A 1990 AARS winner, with slightly perfumed, 10-cm (4-inch) blossoms of coral-pink. Very resistant to mildew and rust.
- 'Redgold'. Rich, 10-cm (4-inch), gold-edged blossoms keep coming all summer. Good perfume and looks exceptional as a cut flower. AARS winner in 1971.
- 'Sexy Rexy'. This cheekily named rose has neat, light to mid-pink blooms. The 9-cm (3½-inch) blossoms have a slight perfume and you can count on constant flower production. Looks like a small hybrid tea. Very good disease resistance.
- 'Sunsprite' (also called 'Friesia' and 'Korresia'). This very fragrant, deep yellow, double rose blooms prolifically in attractive clusters, and the flowers last a long time. Excellent disease resistance, glossy foliage, but needs good winter protection.
- 'Trumpeter'. I've found this to be one of the best performers under heat stress. Scarlet-orange, 9-cm (3½-inch) blooms have only a slight perfume, but it has dark green, glossy foliage, and good disease resistance.

Personal picks—grandifloras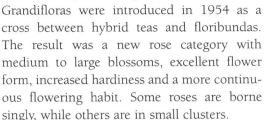

Grandifloras were introduced in 1954 as a cross between hybrid teas and floribundas. The result was a new rose category with medium to large blossoms, excellent flower form, increased hardiness and a more continuous flowering habit. Some roses are borne singly, while others are in small clusters.
- 'Love'. This 1980 AARS winner was the very best of the trio dubbed 'Love', 'Honor', and 'Cherish'. The 9-cm (3½-inch), lightly scented blossoms are an interesting combination of scarlet with a silvery white reverse. It has good mildew resistance and makes a great cut flower.
- 'Queen Elizabeth'. This 1955 AARS winner is simply outstanding. It's a wonderful pale pink with a slight perfume and its flower production is prolific. It grows to about 1.8 m (6 feet) and is hardy and disease resistant.
- 'Tournament of Roses'. The coral-pink blossoms of this 1989 AARS winner lighten with maturity and create a lovely bicolour look. 10-cm (4-inch), fully double, lightly perfumed blooms are borne in long clusters. Dark green, glossy foliage.

Personal picks—polyanthas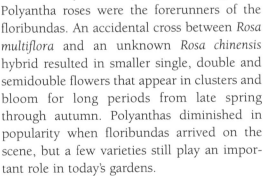

Polyantha roses were the forerunners of the floribundas. An accidental cross between *Rosa multiflora* and an unknown *Rosa chinensis* hybrid resulted in smaller single, double and semidouble flowers that appear in clusters and bloom for long periods from late spring through autumn. Polyanthas diminished in popularity when floribundas arrived on the scene, but a few varieties still play an important role in today's gardens.
- 'Cécile Brünner'. Also known as "The Sweetheart Rose," the small, elegant, shell-pink blossoms repeat freely through the summer on lengthy stems. It grows to about 1.2 m (4 feet) and produces a slight, but distinctive, perfume.

ROSES 109

- 'China Doll'. A popular tiny rose that grows only 50 cm (20 inches) tall, but is absolutely covered with semidouble, deep pink blossoms from late spring until winter. Well suited to containers with winter protection.
- 'The Fairy'. It was first introduced in 1932 and is still a great choice, with its large clusters of tiny, slightly perfumed pink roses. The dark, glossy, fernlike foliage has good disease resistance.

David Austin Roses

David Austin's line of roses first came to prominence in the 1970s. It originated from crosses between old roses and modern hybrid teas and floribundas and combines the best characteristics of each. These roses offer the wide range of colours and recurring blossoms of modern roses while retaining the fragrance and flower form of old roses. Over the years I have grown to be quite a fan, because they have proven to be good garden roses due to their hardiness, relative disease tolerance and versatility. They take about two or three years in the ground to reach their full potential. I recommend planting them in groups of threes, and allowing them to grow together to form a dense shrub that provides a continuous display of colour and fragrance.

Most of the varieties bud up early and keep on blooming throughout the summer. Most are self-cleaning—the blossoms fade, drop their heads and shrivel without diminishing the beauty of the overall plant. Their colour tends to be the best in the cool weather of spring and fall, fading during periods of extended heat. In my experience I would place them in zone 6, with some protection in bad winters.

Personal picks— David Austin roses

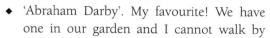

- 'Abraham Darby'. My favourite! We have one in our garden and I cannot walk by without stopping to smell this rose. It's a soft peachy pink, maturing almost to a pale yellow. It grows just under 1.8 m (6 feet) high and can be used against a wall.
- 'Evelyn'. This rose has the strongest perfume of all of David Austin's English roses. The blossoms are apricot with just a hint of pink, turning later to a true pink. Makes a good, strong 1.5-m (5-foot) shrub.
- 'Gertrude Jekyll'. Large, deep pink blossoms with a powerful damask perfume. Good vigour, with bushes up to 1.5 m (5 feet).
- 'Graham Thomas'. The best-selling David Austin rose, with soft yellow blossoms and a tea rose fragrance. A bit shorter, growing to about 1.2 m (4 feet).
- 'Heritage'. David Austin's personal favourite. It's a soft blush pink and has a strong old rose perfume with just a hint of lemon. A good, continuous flowering plant that grows to about 1.2 m (4 feet).
- 'Perdita'. Very hardy and disease resistant, and blooms recurrently through summer. Large, delicate, old-fashioned flowers in apricot-pink tones. It has a lovely fragrance and thrives in poor soil.

Shrub Roses

Shrub roses have been slow to achieve widespread popularity in North America, but they are catching on. The name is a bit of a catch-all for a diverse range of roses, and is simply a rose that you treat like any other flowering shrub. It can go anywhere in the garden as long as it's sunny; like other roses, shrub roses need a minimum of four to six hours of sun. They make attractive hedges and work well in borders.

The best candidate is a rose that is self-cleaning (drops blossoms quickly after blooming), blooms continuously during the growing season, and will outgrow disease problems, such as rust, black spot or mildew. Most modern shrub roses are hardy to zone 6 with some protection. Some, like 'Bonica', will tolerate colder zones, but it's best to trial them first. The size may vary, depending on soil and site conditions.

- 'Ballerina'. This plant produces a mass of single, dainty, light pink flowers with a white centre and yellow anthers. Tiny orange fruits persist into the fall. Can be used in containers with winter protection.
- 'Bonica'. This is the only shrub rose to make the AARS selections, and deservedly so. This double, pink, lightly fragrant rose just never quits blooming from June until frost. It is very disease resistant and has attractive orange hips in the fall.
- 'Carefree Delight'. Growing to only 1 m (3 feet) tall, it produces huge quantities of single, light pink flowers with yellow stamens all through the growing season. Good perfume.
- 'Lady of the Dawn'. It's hard not to love this ruffled, semidouble creamy pink rose with yellow stamens. Dark green foliage sets it off nicely. Grows to about 1.2 m (4 feet).
- 'Snow Owl'. Very compact and rugged, semidouble, white hybrid rugosa that is strongly fragrant. 8-cm (3-inch) blossoms.

Personal picks — *Rosa rugosa*

A popular group of shrub roses is the *R. rugosa* hybrids. They generally have crinkled leaves that are disease resistant and a good bushy habit that makes them great for hedges. Most have spectacular fall hips.

- 'Alba'. Zone 3. Enormous, fragrant, single white blossoms bloom in spring and very sparsely later. Huge, orange red hips and great autumn foliage colour provide interest in fall and winter.
- 'Buffalo Gal'. Low-growing, hybrid rugosa with excellent disease resistance. Semidouble, lavender-pink blossoms have outstanding fragrance. Recurrent blooming spring through fall.
- 'Frau Dagmar Hartopp', or 'Frau Dagmar Hastrup'. A single, with silvery pink blossoms and dominant golden stamens. Blooms profusely and produces large crimson hips in late summer. Grows to 1 m (3 feet).
- 'Hansa'. Zone 3. This shrub rose is one I've grown as a hedge. Its deep purplish red, very fragrant flowers bloom mostly in spring, but sporadically throughout summer and fall. It is disease resistant and tolerant of drought, ocean spray and shade.
- 'Roseraie de l'Hay'. Zone 3. Semidouble, strongly scented variety with wine-red blooms. Good fall foliage, and blooms constantly. Tolerates some shade and makes a nice hedge.
- 'Sir Thomas Lipton'. Zone 3. Masses of fragrant, white, double blooms in spring and again in late summer. Produces hips, and is gaining popular as security fencing because of its thorns.
- 'Thérèse Bugnet'. Zone 3. The huge, double, pink-crimson, very fragrant blooms repeat several times over the season. Pink-toned foliage is very attractive. It tolerates shade, but is susceptible to mildew problems.

Climbers and Ramblers

Climbing roses are instrumental in creating the "dreamscape" so many people strive for in their gardens. They can clothe an arbour or trellis, run along a fence, climb strategically placed pillars or even cover up an old building. Most old climbers have evolved as crosses of hybrid teas and noisettes (a hybrid developed in the late 18th century) or as sports (mutations) from hybrid perpetuals.

Ramblers differ from true climbers in that they have up to 6-m (20-foot), slender canes with small clusters of flowers less than 5 cm (2 inches) across. They generally bloom only once, but in such profusion that they more than make up for their one-shot appearance. They are hardy, but many varieties can mildew, especially if air circulation is poor.

Modern climbers cover a wide range of roses that are grouped together because of their similar growth habits rather than other characteristics. They are all hybrids and their flowers and foliage vary considerably.

Personal picks—old climbers

- 'Climbing Cécile Brünner'. A vigorous sport of the 'Cecile Brünner' bush variety, with beautiful, slightly fragrant, freely produced, shell-pink flowers. Will grow up to 7.5 m (25 feet) high and spread to 6 m (20 feet). Tolerates some shade.
- 'Climbing Souvenir de la Malmaison'. This is one of the most attractive bourbons, producing fragrant, flat, quartered blossoms of blush white shaded with pink. It blooms in summer, with some repeat. Grows to about 3.5 m (12 feet). Prefers sunny climates.
- 'New Dawn'. A climber that produces clusters of semidouble, silver-pink flowers with a slight fragrance. Blooms continuously, tolerates some shade and grows to over 3 m (10 feet). Foliage is grey-green.
- 'Paul's Scarlet Climber'. This cluster-flowering climber grows to about 3 m (10 feet). Bears scarlet blooms in summer, tolerates some shade and has a slight perfume.

Personal picks— modern climbers

- 'Aloha'. This 'New Dawn' cross produces good-sized, rose-pink blossoms with a deep pink reverse. It blooms all summer and grows to 3 m (10 feet) with very lush, dark green leaves. Good for cutting.
- 'Compassion'. A large-flowered climber with nicely shaped buds opening to double flowers in a soft, apricot-pink blush. It's a repeat bloomer with a pleasing fragrance, growing over 3 m (10 feet).
- 'Constance Spry'. A highly prized shrubby climber with large, cupped, rich pink, very fragrant blossoms. Stretches up to 6 m (20 feet) and blooms only in summer. Tolerates shade.
- 'Don Juan'. Overshadowed by its similar cousin, 'Blaze', 'Don Juan' has fragrant, dark red blossoms that bloom over a long period. The glossy foliage gives it good disease resistance. It's great as a pillar, growing up to 2.4 m (8 feet).
- 'Golden Showers'. Still one of the best yellows with deep golden, lightly fragrant blossoms that fade to soft yellow. Blooms constantly all summer with a backdrop of glossy foliage. Grows to 3 m (10 feet) and is great on a pillar.
- 'Royal Sunset'. Large, double, 12-cm (4½-inch) apricot-pink blossoms with a citrus scent bloom all season. Grows to 3 m (10 feet) with disease resistant, glossy foliage.

Groundcover Roses

Groundcover roses are perfect for smothering a bank or garden bed in low-lying flowers. They look great drooping over rock walls or down the edge of a large ceramic pot. Use the varieties hardy enough for your region and plant them close together to control weeds; you really want a spot where they can take over. Groundcover roses have intricate stems and some foliage remains in mild winters (zone 6 and above), so they look good in winter and provide impressive colour during the flowering season. Most are disease resistant and need only minimal pruning.

- 'Baby Blanket'. Double, light pink blossoms persist all summer and well into fall on this vigorous, disease-resistant rose. It's a rugged little performer growing to about 1 m (3 feet) tall and spreading to 1.5 m (5 feet).
- 'Flower Carpet'. This has been a big hit worldwide, with its abundance of small,

semidouble, rose-pink blossoms throughout the growing season. It has outstanding hardiness and disease resistance and gets even better after a couple of years in the ground. The white and apple-blossom varieties seem to have similar characteristics. Grows to about 75 cm (2½ feet) high and spreads to about 1.2 m (4 feet).

- 'Jeeper's Creepers'. Striking single white blossoms cascade from a plant 1.5 m (5 feet) across and 75 cm (2½ feet) high. Blooms continuously.
- 'Red Ribbons'. Very strong grower with good disease resistance and brilliant red, semidouble flowers all summer. Grows to about 75 cm (2½ feet) high and spreads to about 1.2 m (4 feet).

Miniature Roses

Miniature roses are the tiniest of all the roses and are growing in popularity, spurred on by the steady stream of wonderful new varieties each year. It's a bit sketchy, but today's miniatures are probably derived from the China rose, *Rosa chinensis* 'Minima'. In the early 1900s an exceptional variety called 'Rouletii' became popular. Since that time, hundreds of new varieties have been introduced, with a myriad of growth habits, flower colours and sizes.

Ideal miniature roses should resemble larger varieties in miniature form, display attractive foliage, and bloom continuously over a long period of time, with minimal disease problems. If you're buying for your garden, make sure you don't pick one of the many greenhouse varieties, which don't stand up to outside conditions. Ask before you buy.

The miniatures make great container plants, and a few are even grown in hanging baskets. Miniature climbers that grow to 1.5 to 1.8 m (5 to 6 feet) are very attractive in the right spot. In the garden, the good, self-cleaning varieties provide a wonderful show from June through September. Some miniatures, such as 'Nozomi', a beautiful double white, cross over into the groundcover category and are spectacular draped over rocks or low walls.

Miniatures should be planted from the beginning of April to late summer. They prefer a sunny location and good drainage. They need moisture, especially during hot, dry weather.

I'm very impressed with the spectacular large-flowering miniatures, called the Sunblaze Series, from the Meilland rose breeders in France. They are, however, prone to mildew and black spot under poor growing conditions.

Personal picks— large-flowered miniatures

- 'Apricot Sunblaze'. Fragrant, orange-red flowers bloom continuously. Grows to 40 cm (16 inches).
- 'Orange Sunblaze'. Vibrant, orange-red, slightly scented flowers bloom summer to autumn. Good disease resistance.
- 'Pink Meillandina', or 'Pink Sunblaze'. Clear pink flowers bloom reliably in all weather. Grows to 40 cm (16 inches).
- 'Red Rascal'. One of the most floriferous miniature shrub roses, with nonstop, 6-cm (2½-inch), self-cleaning red blossoms. The plant is compact, growing to 1 m (3 feet) with miniature, mildew-resistant foliage.

Personal picks— miniature climbers

- 'Laura Ford', or 'King Tut'. 5-cm (2-inch), golden yellow blossoms with a distinctive pink tinge adorn this 2-m (7-foot), continuously blooming climber. It has won awards in Europe.
- 'Rosalie Coral', or 'Rocketeer'. Continuously produces bright, 5-cm (2-inch), coral-orange blooms. Grows up to 2 m (7 feet). Glossy, dark green foliage creates a nice contrast.

Rose Care

Most folks believe spring is the only time to plant roses, but this is not so. If you plant in the fall, you'll give your roses a head start—and you'll see better, more abundant flowers in the first season. Fall is also the time to move existing roses to a new location in zone 6 and above, but be sure to follow the tips below on drainage and soil preparation. And don't forget to winterize your plants—it will pay off with big dividends in the spring.

Soil preparation and site selection

This is crucial for success with roses. It makes a huge difference in the ability of the plant to survive and its rate of growth after planting. The planting hole should be at least 45 cm (18 inches) wide and 45 cm (18 inches) deep. Into this hole place well-rotted manures, bark mulch and sand to provide nutrients and ensure good drainage. If the roses have been budded, plant with the bud graft just below the soil level for added winter protection.

praying

Using a dormant spray in the winter helps control fungal and insect infestations. In my experience, this controls 80 percent of insect and disease problems and leaves the rose stems clean, healthy and shiny-looking.

The spraying should be done in late fall (mid-November in the coastal region), again at the end of January, and finally at the end of February. Use a combination spray of lime sulphur and dormant oil. Check with your local garden centre for more information.

Roses prefer slightly acidic to neutral soils, but they also love well-rotted manures. Mushroom compost is an excellent choice, and 10 to 15 cm (4 to 6 inches) should be applied in the spring. Once they start growing, feed them with a rose food, such as 6-9-18 or 10-14-21. Repeat every eight weeks until mid-August. Landscape fabric around the base of the plants is a good way to keep weeds away, or weed by hand or use chemical control. However you do it, keep those nutrient-robbing weeds away from the base of the rose.

A minimum of four to six hours of sun a day is required for these sun lovers, but some will make do with partial shade. Good air circulation helps prevent mildew and black spot, but don't expose roses to a lot of wind.

Winterizing

This is essential if you live in areas with harsh winters, but even in moderate coastal climates most plants will benefit from winterizing. How much winter protection your roses need is determined by two factors: the hardiness of the plant and your climate zone.

If you plant nonhardy roses in an area with a harsh climate, you will have your work cut out for you. Lots of mulch in the form of bark mulch, straw or other insulating material will be needed, and even then the plant may not survive. Always match your plant selection to local conditions.

In late fall, cut bush rose stems back to about 1 m (3 feet) to tidy them up, using sharp, clean shears. Remember to cut at a 45° angle. Next, remove any obviously dead wood, which is the ideal winter hangout for insects and disease. Pruning paint on the cuts will keep borers from getting inside. Pick off any remaining leaves, and rake up all decomposing material near the base of the plant.

Climbing roses are treated differently when it comes to winterizing. Choose four or five of the youngest canes and prune them back to about 1.8 to 2.4 m (6 to 8 feet). Cut everything else back to the bud union. Tie the remaining canes to a trellis or arbour to prevent winter winds from thrashing them about.

Finally, generously cover the bud union with soil, bark mulch or other material to protect it from winter cold.

Even though they appear to be dormant, roots are still developing during winter in milder zones (6 and above). Use a winterizing fertilizer, such as 6-9-18, or 4-10-10, applied in a band directly below the outermost branches. Once this is done, cover the same area with a 30-cm (1-foot) layer of bark mulch or garden compost. Be sure the bud union is covered, to ensure the survival of the plant in even the coldest winter.

Spring pruning

Pruning most bush roses is a relatively simple, but crucial, task. Floribundas, hybrid teas and grandifloras are all pruned in a similar manner. Wait until after the last frost in the spring, even if the plant is beginning to bud. Before pruning, stand back and decide on the number of flowers you want and the height desired when it's blooming. The harder you prune, the more compact the plant will be, leading to fewer flowers of a higher quality. Every few years it's a good idea to be a bit ruthless and prune back hard to force a new set of buds to develop from the bottom.

Prune bush roses back to about 30 to 38 cm (12 to 15 inches). Select three to five of the healthiest, thickest stems and cut out everything else to a point flush with the main stems as close to the central union as possible. The remaining stems should radiate out from the centre of the plant. In the case of floribundas, it is not always possible to find really thick stems, so just concentrate on removing the weaker-looking ones and create a framework of the best ones.

The pruning of shrub roses depends on the variety, but as a rule of thumb, clip out dead, diseased and spindly wood every year. Keep them in check according to the available space in the landscape. If you have lots of room, leave them to grow into their natural size and shape.

Climbers and ramblers are generally pruned or trained to cover their framing structure. Leave the healthiest four or five new canes to follow the shape of the trellis or

Insect and disease control

Crown borers can be a serious problem. These insects burrow into the stems, and down into the plant's base where they lay eggs, which hatch and chew the soft branch tissue, causing stem dieback. Fortunately, it's simple to prevent this from happening, by sealing up the cut ends with pruning paint, and by using dormant sprays in winter.

Aphids can be controlled with a systemic insecticide. Organic pesticides are also available. To keep mildew in check, alternate sprayings of two fungicides and garden sulphur every 10 days. Alternating the fungicides prevents the mildew from building up a resistance to any one of the three products. If you wish to use these sprays, consult a licensed pesticide dispenser at a garden centre for more information.

arbour. These may be left for several years and lightly pruned to produce side shoots, which can in turn be trained as a source of new wood for future blossoms. Remove dead, diseased and spindly wood each year to keep the framework healthy.

Miniatures should be pruned differently from other types of roses. Instead of removing blossoms after they flower (deadheading), prune the entire plant back after each blooming period. With constant heavy feeding, new branches will develop quickly and the plant will double in size after each pruning.

Always use clean, sharp shears. Teflon-coated blades can be cleaned in a solution of 1 part bleach to 10 parts water, while steel blades can be cleaned in a baking soda and water solution (10 parts water to 1 part baking soda, approximately). Dip your shears in the solution before moving on to the next plant to avoid spreading disease from one plant to another.

Pruning:
It's important to prune roses using a 45° angle cut.

The ideal cut is made about .6 cm (¼ inch) above an outward-facing bud (a swelling) and should be made at a 45° angle sloping away from the bud, to shed water. Cutting closer to the bud may destroy it, while cutting too high will promote dieback, which can be a breeding area for disease and insects. The cutting blade should move from bottom to top as it cuts to avoid tearing.

Vines and Climbers

Climbing and trailing vines have become an integral part of our modern gardens. They soften any landscape and disguise unsightly walls and fences. Ugly chainlink fences can virtually disappear and old rundown buildings can be transformed by a judicious use of these plants.

In today's more crowded environments, vines and climbers can make plain privacy screens more attractive and can shelter back gardens and patios without the dense blocking effect of hedging. Arbours, trellises and pergolas add charm and visual interest, as well as privacy, to the garden. I particularly admire folks with sunny decks who install an overhead framework for grape vines—they have shade in summer, fruit in fall and a lovely network of vines to enjoy in winter.

Climbing structures do not have to be expensive or complicated. I was quite taken with a very simple and effective use of vines I saw at the Royal Horticultural Society's Garden at Wisley in England. They had placed 3-m (10-foot) wooden posts in various locations and trained vines and climbing roses up and around them.

Vines and climbers use several different climbing methods. Some are self-clinging and can climb up trees, walls and fences on their own, without special support. They do this with aerial roots or adhesive tendril tips; an example would be climbing hydrangea. Plants that climb by means of twining stems or leaf-stalks, or curling tendrils, require support other than a flat wall. These climbers include akebia, clematis and passion flower. Another group of climbers has hooked thorns or long stems and must be tied into a tree or wall. All climbers require at least initial support until they are established. A string or bit of fish net might be all that's required to get them going.

Although vines are usually thought of as climbers, the other option is to use them as a groundcover, and some of them accomplish this job quickly and effectively.

Vines have so many applications it's not hard to see why they are popular. But remember that many vines can be invasive and overpowering. Some especially vigorous vines can strangle weak trees. Unless you have the space to allow them to go wild in the landscape, they need regular pruning and some control. Some vines are susceptible to fungal diseases and require spraying. Others, such as many of the honeysuckles, seem to attract aphids. It's always wise to ask knowledgeable plant people in your area to suggest more disease- and insect-resistant varieties.

A Climbing Sampler

Best bets—vines for a shady location

Many of us have at least one side of our home that's shady a good portion of the day where we could use a little privacy or we need to screen or enhance a particular area. Virtually all vines will grow in shade, but to have an attractive display rather than a sprawling mess, choose vines that naturally grow in shady areas.

Most of these vines will tolerate some sun in morning or late afternoon, but try to keep them out of the midday sun to avoid burning, fading or shortening their flowering time because of the heat.

Climbing hydrangea (*Hydrangea petiolaris*). Zone 5. Self-clinging and fast growing (once it's established) up to 18 m (60 feet). Looks fabulous up trees and north-facing walls. Flat heads of white lacecap flowers in June. Very attractive foliage.

Dutchman's pipe (*Aristolochia macrophylla*). Zone 6. Very fast-growing, twining plants that are ideal for covering trees, walls and fences. The unusual-shaped flowers look somewhat like a pipe. Pale green leaves in abundance. Grows to 15 m (50 feet). Needs a little support to start with.

Five-leafed akebia (*Akebia quinata*). Zone 5. An elegant, twining semi-evergreen vine with vigorous but dainty leaves and fragrant small purple flowers in April. Bears narrow, sausage-shaped, purplish fruits, 10 cm (4 inches) long, in summer, although these are rarely produced. Grows to 9 to 12 m (30 to 40 feet).

Persian ivy (*Hedera colchica* 'Dentata-variegata'). Zone 6. Huge, showy, yellow-blotched leaves that really lift and brighten a shady location. Needs some support to get going.

Schizophragma hydrangeoides. Zone 6. Similar to climbing hydrangea, with small white lacecap flowers. Will tolerate sun. Grows to 8 m (25 feet).

Best bets—vines with colourful foliage

A little foliage colour can add interest to an otherwise quiet area. We have rave reviews all summer about the male *Actinidia kolomikta* vines with their green, white and pink leaves. Shady areas are lifted tremendously with the large yellow variegated leaves of Persian ivy. In England, the deep purple foliage of grapes adds warm colouring to those rainy days, and it's hard to beat the absolutely brilliant fall colour of Boston ivy and its cousin, the larger-leafed Virginia creeper.

Boston ivy (*Parthenocissus tricuspidata* 'Veitchii'). Zone 4. Small, three-lobed leaves open red-bronze, then mature to a rich glossy green. Fall colouring is a truly spectacular orange-scarlet.

***Euonymus fortunei* 'Emerald Gaiety'.** Zone 5. Better known as a groundcover, this evergreen plant with white-margined small green leaves looks fabulous year-round. It changes colour through the seasons and has a pink tinge in winter. We've used it on our garden pergolas for years. Needs support to start and is slow-growing.

Golden hop plant (*Humulus lupulus* 'Aureus'). Zone 5. Bears soft yellow leaves if grown in full sun. It is most effective on pergolas, metal spires or tripods. It dies down to the ground in winter, but has interesting seed-like pods. Needs support to start but will grow 4.5 to 6 m (15 to 20 feet) in one season.

Ornamental grape, or claret vine (*Vitis vinifera* 'Purpurea'). Zone 5. Leaves open a claret-red and later turn a deep purple. It's very effective next to a grey- or silver-foliaged shrub, or twining up a golden-foliaged tree.

Ornamental kiwi vine (*Actinidia kolomikta*). Zone 6. Remarkable tricolour variegation of green, turning to white on the tips and finally pink. The variegation is not apparent early in the season, but by June in a sunny location it is quite spectacular. The males have the best colouring, but if the small white flowers of the females are pollinated, they form grape-sized edible fruits in fall. Needs some support at first.

Porcelain ampelopsis (*Ampelopsis brevipendunculata* 'Elegans'). Zone 8. Self-clinging and evergreen in mild areas, it has small, mottled white leaves tinged with pink. It's a small plant well suited to small, protected areas and container growing. It produces tiny, metallic blue berries in late summer. Great in hanging baskets.

Variegated ivies (*Hedera* spp.). Zone 6. Variegated Persian ivy (*Hedera colchica* 'Dentata-variegata') is green with creamy yellow markings. Variegated English ivies include *Hedera helix* 'Gold Heart', which has dark green leaves with golden yellow centres on reddish stems. Needs support to start.

Virginia creeper (*Parthenocissus quinquefolia*). Zone 3. 'Star Showers', a newer variety discovered in Boston, Massachusetts, has foliage that is nothing short of stunning. This self-clinging vine has predominantly white leaves splattered with irregular green markings with a pink tinge. It needs support at first. *P. quinquefolia* 'Englemanni' is a finer-leafed Virginia creeper with incredible orange-scarlet foliage in autumn.

Best bets—flowering vines

Green vines add class, colourful vines add life, but flowering vines provide the pièce de résistance with their fabulous flowers and enticing perfume. One wonderful older gentleman brings me a picture each year of his *Clematis montana* in bloom around his home; he's so proud of it. There's an older home in our community I drive by each May just to see the wisteria. These vines are something to wait for each year.

Clematis. (See the "Clematis" section below for details.)

Climbing hydrangea. (See "Best bets—vines for a shady location" for details.)

Honeysuckle (*Lonicera* varieties). Zone 4. This is the sentimental favourite of many gardeners.

- *L. caprifolium* 'Mandarin', a University of British Columbia introduction, bears huge heads of orange flowers in abundance in May and June and then periodically through the summer. It's a very rapid screening plant with attractive, glossy, dark green foliage. It needs support.
- *L.* × *heckrotti* 'Gold Flame' produces perfumed, red and yellow flowers continuously throughout the summer.
- *L. japonica* varieties are some of the most perfumed. 'Halliana' bears tiny white flowers that change to yellow and bloom all summer with a perfume that hangs in the air.
- *L. periclymenum* 'Belgica' has reddish purple flowers in May, June and again in late summer. They have a heady perfume and are followed by red berries in late summer.
- *L. periclymenum* 'Berries Jubilee', introduced by Monrovia Nursery in California, has fragrant yellow flowers in summer and again in fall. The bonus is bright red berries that appear after the flowers; the blue-green foliage is very attractive.

Silver lace vine (*Polygonum aubertii*). Zone 5. This is perhaps the fastest-growing vine of all, which makes it ideal for a quick privacy screen. Small, light green leaves complement the masses of white, lacy flowers in late spring and again in late summer.

Summer jasmine (*Jasminum officinale*). Zone 7. Deciduous, semiclinging vine with masses of tiny, fragrant white flowers June through September. This is the one that's been growing in English gardens for more than 400 years. Needs a permanent support.

Trumpet creeper (*Campsis × tagliabuana* 'Madame Galen'). Zone 6. A deciduous vine that bears huge, vibrant, orange-red, trumpet-shaped flowers in late summer. It's often called hummingbird vine, but as flowers appear so late in summer, most of the hummingbirds are usually on their way south.

Wild passion flower, or maypop (*Passiflora incarnata*). Zone 6. It needs a very sunny, protected location, and tends to perform best on a south wall. The flowers are composed of a flat circle of petals with a fringed corolla in the centre, from which protrude the showy reproductive parts. The effect is intricate and very striking. Most varieties produce a small orange fruit, filled with edible jellylike pulp. It dies completely to the ground in winter and bounces back in spring.

Winter jasmine (*Jasminum nudiflorum*). Zone 5. Deciduous, semiclinging vine with small leaves in summer. Tiny, trumpetlike flowers appear November through March in mild winters. Needs a permanent support.

Wisteria. Wisterias are some of the fastest-growing and most magnificent vines, with long, finely cut leaves and flowers produced in long panicles in May and June. They perform best in sun.

Of all the species, Chinese wisteria and Japanese wisteria are the most widely grown. Many folks become discouraged because the most commonly produced seedlings and cutting-started plants of Chinese wisteria (*W. sinensis*) take seven years to bloom. The wait, however, is worthwhile. You can tell the species apart by their twining method. Chinese wisteria twines counterclockwise, whereas Japanese wisteria twines clockwise.

Wisteria needs support and should be kept in check with yearly pruning to maintain a good-looking appearance.

- Chinese wisteria (*W. sinensis*) is the most popular of all wisterias, with its prolific deep mauve flowers, up to 30 cm (12 inches) long. The blooms have a sweet perfume. 'Black Dragon' has large, double, dark purple trusses with a slight perfume. 'Cooke's Special' has a massive number of blooms.
- Japanese wisteria (*W. floribunda*) bloom as younger plants. 'Rosea' has large, fragrant, pale rose flowers with purple tips that grow to 40 cm (16 inches). 'Royal Purple' has highly perfumed, abundant blossoms of a light rose colour. 'Snow Showers' has dainty white flowers tinged with lilac that grow up to 60 cm (2 feet) long. They are delightfully perfumed. 'Violacea Plena' has very double violet-blue flowers with a slight perfume.
- 'Issai' is a cultivar of another wisteria, *W. × formosa*. It is one of the finest lilac-blue hybrids and bears masses of short trusses.

Clematis

Clematis, Queen of the Vines, deserves a section all to herself. Without a doubt, these are the most spectacular of all climbers. By carefully selecting your plants, you can have clematis in bloom from late winter until late fall.

Clematis can smother walls, fences and buildings with colour in short order. They are at home on any trellis, arbour or pergola and can climb up deciduous or evergreen trees. A few species, such as *C. heracleifolia* and *C. integrifolia*, are herbaceous perennials that are appropriate in the perennial garden.

All clematis love a cool area for their roots to grow, but do best in full sun or partial shade. They need constant moisture and some nutrients to keep their new growth looking lush and attractive.

There are two groups of clematis: species and hybrids.

Best bets—species clematis

Species clematis number in the hundreds. Most have small flowers and, as a rule of thumb, are much easier to establish than the large-flowering hybrids. Species clematis, *Clematis tangutica,* for example, have beautiful silken seed heads. Some are self-supporting by means of their twining leafstalks, while stronger-growing varieties are more at home running up trees, over fences and walls.

The only pruning needed on species clematis is removing dead or unattractive wood and pruning back vines that have gotten out of hand. Late-summer–flowering species can be pruned hard every spring to keep them a little more neat and tidy.

C. alpina. Zone 5. This is one of the hardiest and easiest species to grow. The beautiful, small flowers have four petals that hang down with small white stamens. Flower colours are usually blue, violet-blue or lavender tones. Blooms are produced on last year's growth in April and May. They perform well in shade and the bonus is silvery seed heads. 'Jacqueline du Pré' is a rosy mauve and very vigorous. 'Odorata' is a mid-blue with a slight perfume. 'Pamela Jackman' is a rich, deep purple. All reach 2 to 3 m (7 to 10 feet) in height.

C. armandii. Zone 8. This tender evergreen clematis has long, leathery leaves with creamy white, highly perfumed flowers in April and May. It needs a warm, sunny wall. 'Snowdrift' has large white blossoms and 'Apple Blossom' has pink buds fading to white. Grows to 6 m (20 feet).

C. cirrhosa* var. *balearica. Zone 7. An evergreen species whose leaves turn bronze in winter. In protected areas with mild winters, they flower in November and continue into March. Flowers are greenish yellow with ruby spots. 'Freckles' is a light cream with red blotches. Reaches 3 m (10 feet).

***C. florida* 'Sieboldii'.** Zone 6. This species from Japan has creamy white flowers with unusual centres of violet stamens. Blooms June through August and reaches 2.4 to 3 m (8 to 10 feet). Prune lightly in early spring and again to tidy up at the end of the flowering season. Heavy pruning will reduce blossoms.

C. macropetala. Zone 5. This hardy species grows to 3 to 4 m (10 to 13 feet) with small, double violet-blue to lavender flowers that form early and keep on blooming until June. Silky, fluffy seed pods turn grey with age. They are best on fences and low walls. 'Blue Bird' has 5-cm (2-inch) lavender-blue flowers in April and May. 'Rosy O'Grady' has 5-cm (2-inch) light pink flowers. 'White Swan' has 5-cm (2-inch) creamy white flowers in April.

C. montana. Zone 6. This vigorous clematis originated in the Himalayas. It will grow in colder areas, but if the whole plant freezes back each year, it will not bloom. It is one of the easiest to grow, often reaching 9 m (30 feet) or more. The varieties with darker foliage look great all season, well after the flowers have finished in May or June. It's a good choice for growing through evergreens. All *C. montana* have a scent, but some are stronger than others. They are one of the best vines for hiding unsightly fences. The species has a profusion of 5-cm (2-inch) flowers that appear in May and June.

- 'Elizabeth' bears 5- to 8-cm (2- to 3-inch), triple, clear pink, vanilla-scented blossoms in May and June.
- 'Freda' has deep cherry-pink blossoms and very bronze foliage.
- 'Odorata' has masses of 5- to 8-cm (2- to 3-inch), sweet-scented, pure white flowers in May and June.
- 'Tetrarose' is a bronze-foliaged variety with larger, 9-cm (3½-inch), lilac-rose flowers. More compact.

- var. *rubens* has bronze-purple new growth and leaves with rosy pink flowers.
- var. *wilsonii* has highly scented cream-coloured flowers that bloom May through July. Very vigorous.

C. montana **var.** *sericea* (**sold as** *C. chrysocoma*). Zone 6. Profuse white and soft pink flowers are produced in May and June and again in late summer on new growth. The yellowy down in the centre of the flower is a lovely contrast. 'Continuity' is a soft pink. 'Spooneri' is a pure white. Reach 6 m (20 feet).

C. tangutica. Zone 5. This is an easy-to-grow plant with golden yellow, nodding lanternlike flowers 4 to 5 cm (1½ to 2 inches) long. Flowers look fabulous among masses of silky seed heads. Blooms June to September. Grows 5 to 6 m (16 to 20 feet). 'Golden Harvest' and 'Gravetye variety' both have 5-cm (2-inch) yellow, nodding flowers.

C. terniflora (*maximowicziana*). Zone 6. Otherwise known as sweet autumn clematis, this plant has small, scented white flowers that bloom profusely in September and October. Blooms best after hot summers. Evergreen in milder areas. Very vigorous, forming quite a dense mass. Grows 6 to 9 m (20 to 30 feet).

C. vitalba. Zone 5. Known as traveller's joy or old man's beard, this is the native clematis of English roadsides. White, 8-cm (3-inch) blossoms with very prominent stamens hang in huge panicles from mid-June through September. The seed heads stay on the plant through winter. Grows to over 15 m (50 feet).

C. viticella. Zone 5. These southern European natives have contributed more to the popularity of clematis than any other species. Flowers up to 8 cm (3 inches) bloom profusely on slender stalks from June through August. They have been in cultivation since the 16th century. Grow up to 3.6 m (12 feet).
- 'Little Nell' has small, eye-catching, bi-colour cream and mauve flowers with beautiful green stamens.
- 'Minuet' has flowers up to 9 cm (3½ inches) that are veined red on a white background.
- 'Polish Spirit' has deep purple flowers up to 10 cm (4 inches).
- 'Venosa Violacea' has 10- to 15-cm (4- to 6-inch) blossoms with purple veins over a white background and almost black stamens. Fabulous.

Best bets— large-flowered varieties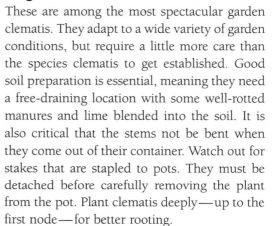

These are among the most spectacular garden clematis. They adapt to a wide variety of garden conditions, but require a little more care than the species clematis to get established. Good soil preparation is essential, meaning they need a free-draining location with some well-rotted manures and lime blended into the soil. It is also critical that the stems not be bent when they come out of their container. Watch out for stakes that are stapled to pots. They must be detached before carefully removing the plant from the pot. Plant clematis deeply—up to the first node—for better rooting.

As a rule of thumb, the large-flowered varieties prefer to have their "heads" in the sun and their roots in a shaded location. Paler varieties tend to bleach out and lose their colour in hot sun, so partial shade is preferable for many pink and bicolour varieties.

These clematis fall into three main groups when it comes to pruning: Group A, group B1 and B2, and group C. For specific information on pruning methods for these groups, see the pruning section at the end of this chapter.

Large-flowered clematis—white
- 'Huldine'. Extremely vigorous plant with 7- to 10-cm (2½- to 4-inch) flowers from July to October. Grows 3.5 to 6 m (11 to 20 feet). Group C.
- 'Madame le Coultre'. Pure white, 15- to 20-cm (6- to 8-inch) blossoms with yellow

stamens that flower June through September. Grows 2.4 to 4 m (8 to 13 feet) Group B2.

Large flowered clematis — purple

- 'Etoile Violette'. Deep purple 10- to 13-cm (4- to 5-inch) blossoms that bloom in profusion from June to September. Very free flowering and vigorous. Grows 2.4 to 4 m (8 to 13 feet). Group C.
- 'Jackmanii'. The first purple hybrid to combine a good, deep purple with large, 10- to 14-cm (4- to 5½-inch) size. Blooms June to August and grows 3.5 to 6 m (11 to 20 feet). Group C.
- 'The President'. Deep purple-blue with a silvery reverse and red-tipped stamens. Blooms June to September with giant, 15- to 20-cm (6- to 8-inch) blossoms. Grows 3 to 4 m (10 to 13 feet). Group B2.

Large-flowered clematis — blue

- 'Blue Ravine'. Large, 20- to 25-cm (8- to 10-inch) flowers of soft violet with dark veining bloom in May, June and September. Grows 2 to 2.5 m (7 to 9 feet). Group B1.
- 'H. F. Young'. Large 15- to 23-cm (6- to 9-inch) mid-blue flowers with creamy white stamens bloom in May, June and September. Grows 2 to 2.5 m (7 to 9 feet). Group B1.
- 'Horn of Plenty'. Bears very large, 15- to 23-cm (6- to 9-inch) rosy purple blossoms from June to September. Beautiful and heavy-producing variety. Grows 2.4 to 3.5 m (8 to 11 feet). Group B2.
- 'Prince Charles'. This smaller-growing plant has pastel blue, 7- to 10-cm (2½- to 4-inch) flowers and blooms continuously June to September. Grows 1.5 to 2 m (5 to 7 feet). Group C.
- 'Ramona'. Lavender-blue, 12- to 18-cm (4½- to 7-inch) flowers bloom vigorously on this old reliable from June to September, and often continue until frost. Grows 2.4 to 4 m (8 to 13 feet). Group B2.
- 'Victoria'. Often known as the blue 'Jackmanii', it blooms from June to September with bluish rose-purple, 10- to 15-cm (4- to 6-inch) blossoms. Grows 3 to 5 m (10 to 17 feet). Group C.

Large-flowered clematis — pink

Plant pink varieties in shade or partial shade, as they lose their colour in hot sun.

- 'Comtesse de Bouchaud'. This is one of the most popular clematis ever grown. Its 10- to 15-cm (4- to 6-inch) mauve-pink flowers bloom profusely June to September. Grows to 2.4 to 4 m (8 to 13 feet). Group B2.
- 'Hagley Hybrid'. Up to 18-cm (7-inch), shell-pink blossoms are borne in early June, with a smaller flush in September. Blooms quickly after pruning during the growing season. Grows 2 to 2.7 m (7 to 9 feet). Group C.
- 'Margaret Hunt'. Dusky pink, 10- to 15-cm (4- to 6-inch) flowers with brown stamens bloom June to September. Grows 2.4 to 4 m (8 to 13 feet). Group C.
- 'Mrs. Spencer Castle'. Produces spectacular, lavender-pink double blossoms in May and June, then single, 11- to 17-cm (4½- to 6½-inch) flowers late in the summer. Grows 2.4 to 3.5 m (8 to 11 feet). Group B1.
- 'Proteus'. This old-timer remains one of the best today. Has rosy 15- to 20-cm (6- to 8-inch) double blooms in May and June, then single blooms through the summer. Grows 2.4 to 3.5 m (8 to 11 feet). Group B1.

Large-flowered clematis — red

- 'Kardynal Wyszynski'. Very deep carmine-red, 12- to 15-cm (4½- to 6-inch) flowers continue June to September. Grows 2.4 to 3.5 m (8 to 11 feet). Group C.

- 'Madame Julia Correvon'. Small, deep wine-red, 7- to 11-cm (2½- to 4½-inch) blossoms from June to September. An amazing bloomer because of its species (*C. viticella*) background. Grows 2.4 to 4 m (8 to 13 feet). Group C.
- 'Niobe'. Very deep ruby-red, almost black petals with striking yellow stamens. Mid-sized 10- to 15-cm (4- to 6-inch) flowers bloom June to September. Grows 2.4 to 3 m (8 to 10 feet). Group B or C.
- 'Ville de Lyon'. Truly a favourite in the reds. Its 10- to 15-cm (4- to 6-inch) blooms have a carmine edge that fades into a lighter interior with large yellow stamens. Light pruning produces larger blooms, which continue June to September. Grows 2.4 to 3.5 m (8 to 11 feet). Group C.

Large-flowered clematis—bicolours

These bicolour varieties do better in shade or partial shade, as they lose their colour in hot sun.
- 'Barbara Jackman'. Interesting purple-blue with deep magenta bar. Good-sized 11- to 18-cm (4½- to 7-inch) blossoms appear in May, June and September. Grows 2 to 2.7 m (7 to 9 feet). Group B1.
- 'Doctor Ruppel'. The 15- to 23-cm (6- to 9-inch), rose-red flowers with a carmine bar bloom May, June and September. Grows 2.4 to 3.5 m (8 to 11 feet). Group B1.
- 'Mrs. N. Thompson'. Deep purple sepals with a striking red bar. Wow! The 10- to 15-cm (4- to 6-inch) blooms flower in May, June and September. Grows 2.4 to 3.5 m (8 to 11 feet). Group B1.
- 'Nelly Moser'. One of the old standbys for over 100 years, it has huge, 17- to 23-cm (6½- to 9-inch), mauve-pink flowers with a carmine stripe. It flowers in May, June and September. Grows 2.4 to 3.5 m (8 to 11 feet). Group B1.

Large-flowered clematis—doubles

As a rule, double-blooming varieties are a wonderful novelty, but do not provide a continuous, strong blossoming habit like the singles.
- 'Countess of Lovelace'. Beautiful bluish lilac with a semidouble flower form. It bears 15- to 23-cm (6- to 9-inch) flowers in June and single flowers in August. Grows 2.4 to 3.5 m (8 to 11 feet). Group B1.
- 'Duchess of Edinburgh'. This spectacular double white has 10- to 15-cm (4- to 6-inch) double blooms in May, June and September. It's over 100 years old and still a winner. Grows 2.4 to 3.5 m (8 to 11 feet). Group B1.
- 'Mrs. P. T. James'. Wonderful 15- to 20-cm (6- to 8-inch), sky-blue double blossoms in June and single blossoms the balance of the summer. Hybridized on Saltspring Island by Mr. P. T. James. Grows 2.4 to 3.5 m (8 to 11 feet). Group B1.
- 'Veronica's Choice'. Double lavender with a grey tinge. The 15-cm (6-inch) blossoms are double in May and June and single in late summer. Grows 3.5 to 4 m (11 to 13 feet). Group B1.
- 'Violet Elizabeth'. Mid-sized, 15-cm (6-inch), elegant, mauve-pink double flowers in May and June. Single flowers later in summer. Grows 2.4 to 3.5 m (8 to 11 feet). Group B1.
- 'Vyvyan Pennell'. The best known of all the double varieties. Flowers are 15 to 20 cm (6 to 8 inches) and violet-blue with reddish overtones. Flowers are double in May and June; single blues and a luminous violet are produced in mid-August. Grows 2.4 to 3.5 m (8 to 11 feet). Group B1.

Pruning clematis

Proper pruning is important, but an improper pruning won't kill the plant, it will only delay flowering. Left unpruned, all types will flower, but the area covered by flowers won't be attractive. Here are the basic rules.

The first February or March after planting, all clematis should be cut back. At this time you should be able to see leaf buds developing as the plant breaks dormancy. Leave two sets of buds on each stem between where you make your cut and soil level. In subsequent years, prune according to which of the following groups the plant fits into.

Group A. These are varieties that flower only on growth produced the previous year. Pruning should consist of cutting out weak or dead stems as soon as they are finished blooming in May or June. Pruning later than June, or severe pruning, will result in fewer blooms the following spring. *C. montana* falls into this group.

Group B1. These are varieties that flower primarily on wood produced the previous season—with a heavy flush of flowers in May and June. This is followed by a second, smaller flush of blooms in September on the current season's growth.

Group B2. These varieties bloom simultaneously on last year's growth and the current season's growth. Group B2 varieties normally bloom from June to September continuously. For pruning purposes, they can be treated either as group B1 or group C. A group C pruning regime every second year is recommended.

Both B1 and B2 plants should be given a light pruning in late February or March. Any weak or dead wood should be removed and a careful spacing of the remaining stems is all that's required. The spacing allows room for the flowers to open pleasingly. A severe pruning will reduce the number of blooms, but will not hurt the plant. If a group B clematis has been neglected for many years, it can be rejuvenated by severely cutting back most of the old growth. Separate and train the new shoots to avoid a tangled mess.

Group C. These varieties bloom only on the current year's growth. Blooms start in early summer and continue through to fall. Plants should be cut back in late February or March to two strong sets of buds on each stem as close to ground level as possible. This will provide a plant with blooms that start near ground level and continue to the top of the plant. The majority of group C clematis, if left unpruned, will start their new growth very close to where last season's growth ended and will quickly grow out of control.

If you want to grow a group C clematis through a tree or have it bloom in an area above its normal blooming height, this characteristic can be used to your advantage. You can prune an established plant at almost any

---

Clematis wilt

The term "clematis wilt" is a catch-all to describe fungal damage that infects the stems of clematis and causes the plant to wilt or die back. The delicate frame of clematis can be easily damaged, providing opportunity for diseases to enter the plant. One way to avoid damage is to make sure your plant is attached securely to its support.

Clematis wilt usually happens just as the flower buds begin to open and can affect the whole plant or just one or two stems. If you see it happening, cut out and destroy the affected stems. Be sure to remove at least 2.5 cm (1 inch) of healthy wood beyond the infected portion. Clematis wilt seldom destroys the whole plant; even if you have to cut the main stem back to the ground, healthy buds will usually be produced at soil level.

height or not prune at all to accomplish your objective. Keep in mind that group C clematis bloom on the current season's growth, so that if treated in an untraditional way, the blooms will be at the top of the plant and a bare stem will appear over the years. This provides an opportunity to plant a lower-growing group B variety to hide the bare stem and extend the blooming season.

Deciduous Trees and Shrubs

Ornamental trees and shrubs have long been the backbone of any landscape. They provide shade, privacy and wildlife habitat, offer winter protection and erosion control, and put oxygen back into the air—especially important in urban areas. By choosing the right deciduous trees or shrubs, your garden can have colour and interest throughout the year.

The boundary between shrubs and trees is somewhat elastic, but in general a tree is a woody plant with one main trunk and a distinct elevated crown. They generally grow to 4.5 m (15 feet) or more. Shrubs can be defined as woody plants, usually multistemmed, that grow to less than 4.5 m (15 feet).

Trees

Trees are a reminder of constancy and growth, and that's why I think it is a wonderful idea to plant a tree on someone's birthday, for an anniversary or to mark any special occasion. Years down the road, the tree will remind you of the event and bring back those memories.

Virtually any size garden can have a tree or two. There are small and medium-sized trees with noninvasive root systems to suit almost any space. But before you drive to your local garden centre to pick up a tree, think carefully about what qualities you want it to bring to your garden. Aside from appropriate size and suitability to the site, there are several things to look for, such as foliage, flowering, and interesting branching and bark.

Choosing a tree

Foliage is one of the main considerations when selecting a deciduous tree. Dense foliage will provide deep shade—perhaps more than you want in a particular area. Some leaves have especially pleasing shapes or move readily in light winds. And, of course, one of the most important aspects of foliage is fall colour.

When choosing flowering trees, select trees that flower at the specific time you want colour in your garden. If you have room for several flowering trees or shrubs, try to spread blooming times throughout the year. But don't forget to consider the appearance of the tree all year long and not just how it looks at flowering time. Remember that the bloomtime is relatively short, and other characteristics are also important.

Don't think that because a deciduous tree loses its leaves in winter, it can't make a statement in the winter garden. Many trees have distinctive branch forms or colourful bark which are an asset year-round, but are particularly beautiful in winter when fully revealed. When considering a tree, also think about companion planting to enhance its full potential in the landscape.

Once you've looked at growth habit, foliage and flowering times, use the following list to check out the other characteristics of the tree you have in mind.

Disease resistance and pest tolerance. A tree may be beautiful, but have very low resistance to pests and diseases. If it's going to be a constant battle with pest sprays and disease controls, it may not be worth it.

Root system. Find out if it has a vigorous root system that might cause problems with drains or other plants. Most trees and shrubs can be planted in a landscape with other types of plants. Those with invasive roots, however, should be avoided in a small garden. Examples include willows, poplars, fast-growing shade trees, forest trees such as the giant cedar, spruce, fir and fast-growing evergreens, such as cypress.

Size. Determine what the mature height will be, and how quickly it grows. It may completely outgrow your landscape in a few years.

Bad habits. Some trees have bad odours, phytotoxic leaves (harmful to other plants), messy fruits or sappy discharges.

Soil preferences. Make sure the tree can tolerate your soil conditions. You can amend your soil a certain amount, but it's easier to plant a tree that is comfortable in your growing conditions.

Hardiness and water preferences. Find out the range of heat and cold the tree will tolerate. For instance, some trees prefer a cold winter period and will not tolerate a mild, rainy winter.

Compatibility with other plants. Some trees will not tolerate underplanting or are toxic to plants around them.

Wind resistance. If you live in a windy area, determine if the roots are strong enough to adequately anchor the tree when it is mature or if it has brittle branches that could snap in a high wind.

Care and feeding

Like all plants, various trees have specific needs for water, nutrients, soil conditions and sunlight. When looking at your tree, consider the parts you don't see as well as those you do. About 80 to 90 percent of the nutrient-absorbing roots of most trees are in the top 25 to 30 cm (10 to 12 inches) of the soil. Most of the root system is concentrated around the drip line, a circular area equivalent to the edge of the largest branches. You have to consider the growing conditions and nutrients in the area taken up by the root system when you are planting and caring for trees in your landscape.

The best time to match the tree to your particular conditions is before you plant. To do this, dig a test hole at your planting site. Does

---

No such thing as a "dwarf"

Forget about the term "dwarf" when referring to trees and shrubs. Instead, think in terms of compact, or slow-growing. All trees and shrubs will keep on growing over time, so don't assume a plant will retain a certain size if someone refers to it as a dwarf. It won't. Slow-growing trees look good in your landscape for longer with less maintenance, but still need pruning to enhance both their appearance and longevity in your garden.

the area drain well and have light, sandy soil or is it full of clay, with water resting just below the surface? This is vital information, because some trees simply refuse to grow in certain soil conditions. Find out ahead of time and save yourself grief. Most trees need good soil drainage to a depth of 1 m (3 feet) or more and exposure to at least four to six hours of sunshine a day.

It's rare that existing soil will present perfect growing conditions. If your soil is heavy, add a mix of sand and fir or hemlock bark mulch. It loosens up heavy soil and allows roots to thrive. If the soil is very fast draining (lots of gravel and/or sand), add well-rotted compost or peat moss to retain moisture. These amendments should be made to a wide area—not just the hole where the tree will be planted.

To get a newly planted tree off to a solid start, use a liquid root-boosting fertilizer as well as some fine bone meal. Both will assist in the formation of healthy roots. If you are amending your soil with manure, make sure it is well composted; if it is too fresh, it can burn the fine roots.

All trees need to be pruned on a regular basis. After many years a tree can grow too large and become a landscape liability instead of an asset. But by keeping it pruned, it can look great next year and ten years from now. There is a bit of a learning curve in terms of when to prune what and how to tackle it, but with a little curiosity and common sense you can become a proficient pruner. As a rule of thumb, major pruning on deciduous ornamental and shade trees should be done when they are dormant and the sap is not flowing. Lighter pruning for shape on most broadleaf and flowering shade trees can be done after flowering. Enjoy the blossoms, and when the flowers start to fade, you can cut them back. It's a good idea to consult one of the many publications on pruning before you tackle pruning on a large scale.

Shade trees

Shade trees don't seem to be as common as they used to be in the yards of newer homes. It's a shame, because these tall, stately trees are beneficial in many ways, and don't deserve the bad reputation they have gained in some

Trees for containers

If you're planting a tree in a container, don't make the mistake of squeezing it into something too small. A half whiskey barrel or equivalent-sized container is a good choice: it is big enough to hold moisture in summer and has enough mass to protect the roots from freezing in winter. Prune the lower branches of the tree and plant groundcovers, bulbs or other small plants at its base.

Use a sterilized soil mix in containers and add fine fir or hemlock bark mulch to ensure good drainage.

There are many trees that will adapt to living in containers. Even young shade trees can make great container plants. The size they reach depends on nutrients as well as how much room the roots have. Once the roots hit the edge of the pot and the tree becomes rootbound, growth will be inhibited unless you add nutrients and soil-building material.

quarters. The common misconceptions are that they grow out of control, drop tons of leaves, have destructive root systems, and are prone to insect and disease problems.

The advantages of planting shade trees far outnumber the potential problems. For instance, they provide an effective, inexpensive way to keep your house cool in the summer. One average-sized shade tree has a cooling effect equivalent to four household air conditioners running 12 hours a day. Placed in the right location, it can mean reducing your indoor temperature by many degrees—all without a bill from the power company at the end of the month. And the bonus of a deciduous tree is that it loses its leaves, allowing more light to enter your home in the darker winter months.

All too often I see a house surrounded by a few low evergreen shrubs. This does little to create atmosphere and is a reminder that most of our home landscapes are too flat. Shade trees help create a feeling of depth. No matter what style of home you have, there's a tree shape to complement it. A stately oak or the disease-tolerant sweet gum complements larger homes. Many varieties of birch are also ideal, with their standard pyramid forms. Tall and slender trees like hornbeam or aspen complement smaller homes or tall, narrow houses and condominiums.

The many varieties of maple are perhaps some of the best oval or rounded shade trees to use in the landscape. Redleaf Norway maple 'Crimson King', which is a beautiful burgundy-red all season, or red maple, which turns brilliant crimson in the fall, are two of the best. In colder areas, sugar maples and silver maples are fine specimens that are quite fast-growing.

Speaking of fast-growing, the London plane tree is certainly among the most vigorous. It's also excellent for large city plantings because it tolerates pollution. These are only a few of the many fine shade trees that can enhance any landscape.

Colour is another important factor—fresh, bright green spring growth and fall shades of yellow and red are a bonus, even before you consider the more exotic foliage trees. For summer colour, for example, there is the white-and-green variegated Norway maple, copper beech, golden robinia, or the chartreuse sunburst locust. Fall-coloured specimens are too numerous to mention.

Although you do have to rake up some leaves in the fall with deciduous trees, consider the cleaning job they've done for you the rest of the year. The leaves collect dust all summer, dust that would otherwise end up in your home. They also help to replenish oxygen and reduce carbon monoxide, as well as provide habitat for birds, which then eat insect pests. I could go on about noise reduction and other benefits, but if you're not sold by now you never will be. (Did I mention how much fun kids have climbing in trees or playing in their shade?)

If you're not looking for a huge shade tree, consider something that's slow-growing or one of the more compact varieties (remember, there are no dwarf trees). If insects are holding you back, then think gingko, sweet gum and tulip tree—they're all relatively insect free. If you're worried about invasive roots, seek out a tree with a fibrous root system; they are quite safe, even in tiny lots.

Best bets—large shade trees

Planting a large tree in your garden is an important decision. Even though it's an environmentally friendly thing to do, large trees will end up as 50-m (160-foot) giants and outlive us by many years. Always consider overhead wires, underground services, nearby pavement and the area it will eventually shade.

Katsura (*Cercidiphyllum japonicum*). Zone 5. A medium-sized tree to 5 to 20 m (16 to 65 feet). Its small, round leaves open a reddish colour and turn green with grey undersides. It has fabulous yellow to scarlet colours in fall and is very pest resistant.

London plane tree (*Platanus acerifolia*). Zone 5. I always recommend this fast-growing tree for people who want shade and want it soon. It tolerates pollution and has large, maplelike leaves that yellow in fall. Interesting burrlike fruits hang on the tree in winter. Grey bark on older trees peels and flakes for a most attractive winter appearance.

Maidenhair tree (*Ginkgo biloba*). Zone 5. This is one of the oldest known trees. Fan-shaped green leaves turn bright golden in fall. Female trees produce a nutlike fruit, which can be messy, so make sure you get a male. Pest free, they thrive on pollution and grow slowly to 6 to 20 m (20 to 65 feet). 'Autumn Gold' is one of the best varieties.

Pin oak (*Quercus palustris*). Zone 4. One of my personal favourites, it has a pyramidal form, height of 6 to 20 m (20 to 65 feet), and graceful, slender branches that droop slightly on the outside. Shiny, deeply lobed green leaves turn rich scarlet, then a true brown, and stay on the tree to rustle in the winter winds.

Red maple (*Acer rubrum*). Zone 5. One of the best selections is 'Red Sunset', which quickly grows to 6 to 15 m (20 to 50 feet) and holds its brilliant red fall colour longer than most other varieties. For narrow spaces, try 'Scarlet Sentinel', a columnar form that tolerates summer heat better and is one zone hardier.

Sweet gum (*Liquidambar styraciflua*). Zone 5. This disease-free tree has medium-sized, maplelike foliage and interesting corky bark. It grows moderately fast, reaching 7 to 20 m (23 to 65 feet), and has incredible red, yellow and orange fall colour.

Tricolour beech (*Fagus sylvatica* 'Purpurea', sold as 'Tricolor'). Zone 4. This medium-sized tree, 6 to 12 m (20 to 40 feet), has purplish green leaves edged with white and pink. On a sunny day, looking up under the tree is like looking at a kaleidoscope. This slow-growing tree is also suitable for containers.

Tulip tree (*Liriodendron tulipifera*). Zone 5. This very clean tree is fast-growing, has tulip-shaped leaves and after 10 years has yellow-green, tuliplike flowers with orange centres in June. Great lemon-yellow fall colour. Grows 8 to 24 m (26 to 80 feet). Variegated leaf form in 'Majestic Beauty' (*L. tulipifera* 'Aureo-marginata') is outstanding.

Best bets—small shade trees

Bloodgood maple (*Acer palmatum* 'Bloodgood'). Zone 5. This maple has a rich, deep purple-red foliage that holds its colour in summer, and becomes more scarlet in fall. It's a little smaller at 5 m (16 feet), so works well in smaller spaces.

Compact Japanese maple (*Acer palmatum* 'Fireglow'). Zone 5. For folks with very limited space who want a redleaf Japanese maple, this is the one. It has the same characteristics as the bloodgood maple, but is a very slow grower.

Coral bark maple (*Acer palmatum* 'Sango kaku' or 'Senkaki'). Zone 5. This tree is breathtaking in fall and winter, with vibrant yellow leaves in fall and intense, coral-coloured stems which are beautiful in their own right. It needs good drainage and will not tolerate bitterly cold winter winds. Grows 5 to 6 m (16 to 20 feet).

Green Japanese maple (*Acer palmatum*). Zone 5. I love multiple-stem varieties for their graceful shape and usefulness as screens. Bright fall colours of yellow, orange and red. Grows relatively quickly to 4 to 7 m (13 to 23 feet). Most people tend to pick the red Japanese maples, but I think the green is fabulous.

Japanese red maple (*Acer palmatum* 'Atropurpureum'). Zone 5. The most popular red maple, with leaves that start out a rich crimson, fade to a green-burgundy in the heat of summer, then blaze crimson as the temperature drops in September. Grows quite quickly to 7 m (23 feet).

Paperbark maple (*Acer griseum***).** Zone 5. This is one of the last maples to leaf out during the spring, but its red, peeling winter bark makes the wait worthwhile. Green leaves turn brilliant red in fall. This tree grows slowly to 6 to 8 m (20 to 26 feet), and is very insect and disease free.

Best bets—flowering trees

These trees add a new dimension to the landscape. Not only do they provide shade and screening, but they display flowers, fruit and even great fall foliage. Not all do well in all climates, so make sure their weather and disease tolerance are suitable to your climate.

A flowering tree can really enhance your landscape if you plant it in the right location. Consider the shape of the tree, the appearance of the bark and the colour of the foliage in spring, summer and fall. If you select a wide-spreading tree like a magnolia or dogwood, it will probably look better as a background tree, rather than a focal point in the centre of your yard. These trees are ideal as a backdrop for lower-growing flowering shrubs such as Japanese azaleas or dwarf rhododendrons. A pink dogwood and late-blooming white azaleas is a classic combination.

A standard tree—one that has at least 2 m (7 feet) of straight stem before branching begins—is well suited to being a focal point in the landscape. The key here is underplanting. (For more on this, see "Companion planting" later in this chapter.)

Don't forget to consider foliage colour when shopping for a flowering tree. Remember, the flowers last only a few weeks, but the leaves hang in there for six months. Look for trees with bright spring colour, good rich summer tones and beautiful autumn shades.

Eastern redbud (*Cercis canadensis* **'Forest Pansy').** Zone 4. Stunning cerise-pink flowers along the stems create a wonderful display. New, heart-shaped leaves open maroon and mature to maroon-green. Needs very well-drained soil and sunny location. Moderate grower, 3 to 6 m (10 to 20 feet). Flowers April to May.

Flowering crabapples (*Malus* **varieties).** Zone 3. Flowering crabapples are magnificent, but in wet areas scab-free varieties are a must. 'Red Jewel' has become one of my favourites because single white blossoms produce large red fruits, about 2 cm (¾ inch) across, which stay in place all winter for a spectacular display. Grows 5 to 6 m (16 to 19 feet). Flowers in May.

Japanese dogwood (*Cornus kousa***) and Chinese dogwood (***Cornus kousa* **var. *chinensis*).** Zone 5. Today there are many new varieties, with varying flower forms and branching habits. Some, like 'Gold Star', even have variegated foliage. 'Satomi' has pink flowers. For overall performance I still like the species. All of these trees will take full sun, but flower longer in partial shade. Their four-petaled, white flowers open in June when most

---

Japanese maples

All Japanese maples need open, porous, well-drained soil to do well, especially in wet climates. Add a generous amount of bark mulch to the planting hole. Planted in wet, heavy soils, roots tend to rot and branches will die back. Prune back dead branches, improve soil drainage, and use fungicides if necessary, to help control fungal and bacterial problems. Another choice, if your soil is poorly drained, is to grow it in a large container or a raised bed.

other flowering trees are finished. They produce big, strawberrylike fruits in the fall. Fall foliage is a spectacular yellow to scarlet. These are very disease-free trees and are well suited to a variety of landscape conditions. Flowers June to July.

Japanese flowering cherries (*Prunus* varieties). Zone 5. These trees put on a spectacular display in spring and are very popular in yards and along boulevards. There are many varieties to choose from, and each has its own distinct form, fall foliage colour and flower type.

- One of my favourites is the winter-flowering cherry (*P. subhirtella* 'Autumnalis'), because it blooms from October through April in zone 6 and warmer. Deep pink single flowers open to soft pink and will freeze

Dogwoods and disease

Dogwoods have a lot to offer but at times they are badly in need of our assistance, especially during cool, wet springs. In fact, some people resist planting these remarkable trees because of fears over disease problems, which can be overcome relatively easily. Crown canker and dogwood leaf blotch are the main culprits. Cankers appear on either the branches or the trunk of the tree near the soil surface. As the canker eats into the tree, the bark peels away, which leads to a slow, sure death.

Canker is best prevented by good tree care. Dogwoods planted in sandy, very well-drained soil seem to fare much better—they tend to harden off more quickly in autumn, tolerate heavy rains better and seem to have a sturdier habit. Planting them in bermed soil or raised beds is also beneficial. Crown canker is often started by the person behind the lawnmower or weedeater, who nicks the bark and opens up a site for infection. Prevent this by hand-weeding or using a safe herbicide near the base of the trees. If you do notice a cankered area, trim the discoloured portion out with a sharp, clean knife and treat the bark with a pruning solution. The earlier you treat it, the better, to prevent the disease from spreading.

Leaf blotch is a fungal disease which shows up as brown spots, blotches and wedge-shaped patches at the leaf tips. As this fungus spreads, leaves begin to drop as though it's autumn. Twigs and branches may then develop cankers and die back. If you notice dead terminal buds on your trees, or quite a few dead branches, chances are your dogwoods are infected. Unchecked, it will spread to larger branches and eventually kill the tree.

If you have a tree with leaf blotch, make sure you rake up and burn or bag the leaves to prevent further infestation. Cut out and destroy dead or diseased twigs and branches. You can help prevent leaf blotch by spraying the tree at bud break with a systemic fungicide. See a qualified dispenser at your local garden centre for information.

If you're planting a new tree, avoid the disease-prone Pacific dogwood. Choose species resistant to leaf blotch: the eastern dogwood (*Cornus florida*) or Japanese dogwood (*Cornus kousa*).

off in cold spells. New buds open in mild weather and this pattern persists all winter into spring. It grows 6 to 7 m (20 to 23 feet) in a rounded form.

- 'Daybreak', also called 'Akebono', has huge, cup-shaped, pink flowers hanging by the thousands to make a spectacular display; it's very popular with landscape architects.
- Of the spring-flowering Japanese cherries, 'Kwanzan' is the most popular, with its huge, dark, double pink blossoms.
- 'Pink Perfection' is similar in many ways; I really like its soft, fully double pink blossoms and overall elegance.
- 'White Goddess' or 'Shirofugen' is another winner, with soft pink buds opening up to pure white hanging clusters. It has good fall colour and new growth is a beautiful copper colour.
- Of the white-flowering Japanese cherries, my favourite is the single white 'Shirotae', or 'Mt. Fuji'. Its long, drooping clusters of fragrant flowers hang from the tree like Christmas baubles.
- If you prefer a double white blossom, 'Shogetsu' is the tree for you; it's a lower, flat-topped tree with almost horizontal spreading branches. Flowers hang in huge clusters—an impressive sight.

Japanese snowbell (*Styrax japonicum*). Zone 5. This is a tree for almost any landscape situation. Dainty, fragrant white flowers hang like bells throughout the tree. The green leaves perch like butterflies on the stems, showing off the flowers. Clean, fast-growing, with fan-shaped branches, it grows 7 to 9 m (23 to 30 feet) and has yellow fall colour. Flowers in June.

Magnolias. Magnolias are being used more and more as small- and medium-sized shade trees. Ideally suited to small gardens, whether growing naturally or trained as standard trees, magnolias provide shade in summer, outstanding flowers in spring and attractive winter buds. Most magnolia varieties—although generally grown as bushy, tall shrubs—can be trained in this fashion.

Magnolias as small shade trees:
- *M. dawsoniana*. Zone 6. Grows to 10 m (33 feet). Large, pale rose flowers suffused with purple on older trees. Very long, 15-cm (6-inch) leaves. Lightly scented. Older trees produce blooms.
- *M. denudata*. Zone 5. Grows to 9 m (28 feet). Fragrant, pure white, cup-shaped flowers.
- *M.* 'Elizabeth'. Zone 4. Grows to 12 m (39 feet). Hardy and tolerates a wide variety of soil conditions. Soft yellow, fragrant flowers.
- *M.* 'Galaxy'. Zone 5. Grows to 12 m (40 feet). Dark red-purple buds open into a lighter shade of flower. Tree has a nice form and blooms have a slight perfume.
- *M. kobus*. Zone 4. Grows to 10 to 24 m (33 to 80 feet). One of the hardier magnolias with white, starlike, fragrant flowers. One of the most versatile magnolias.
- *M.* 'Yellow Bird'. Zone 4. Grows to 12 m (40 feet). Medium-sized, deep yellow blossoms flower after the leaves are out.

Bush magnolias for smaller spaces:
- *M.* × *loebneri* 'Leonard Messel'. Zone 5. Grows to 2.4 m (8 feet). Similar to 'Royal Star' but with pink flowers.
- *M. sieboldii*. Zone 6. Grows to 2 to 5 m (7 to 17 feet). A spreading magnolia with white, highly fragrant blossoms appearing among the leaves. Produces showy clusters of crimson fruit in fall. Blooms June to August.
- *M. stellata* 'Royal Star'. Zone 5. Grows to 2 to 4 m (7 to 13 feet). Compact shrub with masses of fragrant, white, star-shaped flowers that appear on young plants.
- *M.* 'Susan'. Zone 5. Grows to 2.4 m (8 feet). Compact magnolia with reddish purple, saucerlike blossoms that bloom over a long period.

Single-flowering plum (*Prunus cerasifera* 'Thundercloud'). Zone 5. This is one of the best flowering plums, with masses of soft pink, single flowers in spring and small purple leaves all summer. Grows to 7 m (23 feet) with sparse but delicious edible fruits. Flowers in April.

Sourwood (*Oxydendrum arboreum*). Zone 5. One of my favourites because it is loaded with masses of lily-of-the-valley–like flowers in August when we need flowering trees in the landscape. Bronzy new growth is a nice accent, and glossy green leaves turn a stunning scarlet colour in fall. Loves sun or shade and grows slowly to 3 to 8 m (10 to 26 feet).

Best bets—trees with interesting bark and stems

Foliage and flowers aren't the only things deciduous trees have to offer. Many varieties provide colour and texture with their bark and captivating form with their branches and stems. The contorted branches of some of these trees are sought after by flower arrangers. If you grow them yourself, you can enjoy the contorted stems on the live plant as well as use the cuttings in flower arrangements. Trees with peeling or colourful bark are also good bets for winter enjoyment.

Birchbark cherry (*Prunus serrula*). Zone 6. The intense red bark of this tree is so shiny and smooth it's hard not to reach out and touch it when you walk by. Small white flowers in April. Grows 4.5 to 6 m (15 to 20 feet).

Contorted mulberry (*Morus bombycis* 'Unryu', sometimes sold as *Morus alba* 'Contorta'). Zone 3. Mulberry has a lot going for it—contorted stems, soft yellow flowers and edible fruit. It grows to 3 m (10 feet).

Contorted robinia (*Robinia pseudoacacia* 'Tortuosa'). Zone 3. The contorted stems and 3- to 5-m (10- to 17-foot) height make this a good tree for shade or screening.

Corkscrew, or contorted willow (*Salix matsudana* 'Tortuosa'). Zone 3. These varieties grow in the 5- to 10-m (17- to 33-foot) range. The roots can be invasive, so I suggest root pruning every two years, or containing them in the ground. They also work well in containers. Two hybrids are 'Golden Curls', with yellow stems, and 'Scarlet Curls', often sold as 'Red Curls'. Both have contorted branches; 'Scarlet Curls' has bright red stems in winter. They look fabulous in winter sunshine and their cut branches are stunning in flower displays.

Harry Lauder's walking stick (*Corylus avellana* 'Contorta'). Zone 5. The contorted stems provide nuts and catkins in winter. Grows to just under 3 m (10 feet) maximum. Be sure to cut out any straight suckers to encourage that twisty growth.

Himalayan birch (*Betula utilis* var. *jacquemontii*). Zone 7. This whitest of all birches grows 5 to 9 m (17 to 30 feet).

Paperback maple (*Acer griseum*). (See "Best bets—small shade trees" above.)

Companion planting

Trees make a terrific addition to the landscape, but you might want to consider some extra measures when the branches are bare. Plantings underneath trees not only enhance their winter beauty, but complement their spring, summer and fall displays.

For instance, Japanese cherries underplanted with the evergreen wintercreeper 'Emerald 'n' Gold' is interesting to look at year-round. In the early spring, the rich bronze colour of unfurling cherry leaves contrasts nicely with the fresh new gold growth of wintercreeper. As that foliage matures to a deep gold, the pink cherry blossoms appear, creating a spectacular sight. When the cherry leaves turn green, the mature variegated foliage of the wintercreeper sets it off nicely. And so it goes through the years...

Underplanting is particularly effective with trees that have downward-arching or weeping branches. Our eyes naturally follow the shape of the branches to the ground, so there should be something there to catch the eye. Colourful azaleas, heather, pieris or rhododendrons are all suitable.

Listed below are some evergreen plants that will tolerate some shade and competition for moisture, as well as some herbaceous perennials that are complementary.

Broadleaf evergreen plants:
- bearberry cotoneaster (*Cotoneaster dammeri*)
- bunchberry (*Cornus canadensis*)
- cliff green (*Paxistima canbyi*)
- creeping Oregon grape (*Mahonia repens*)
- English ivy (*Hedera helix*)
- Japanese spurge (*Pachysandra terminalis*)
- kinnikinnick (*Arctostaphylos uva-ursi* 'Vancouver Jade')
- mountain cranberry (*Vaccinium vitis-idaea*)
- periwinkle (*Vinca minor*)
- salal (*Gaultheria shallon*)
- sweet box (*Sarcococca hookeriana* var. *humilis*)
- wintercreeper (*Euonymus* 'Emerald Gaiety')

Perennials:
- *Ajuga reptans* 'Multicolor' (often sold as 'Tricolor') and 'Rainbow'
- beach wormwood (*Artemisia stellerana* 'Silver Brocade')
- bellflower (*Campanula garganica* 'Dickson's Gold')
- *Bergenia* 'Bressingham Ruby'
- coral bells (*Heuchera* varieties, especially bronze ones)
- creeping thyme (*Thymus* varieties)
- dwarf daylilies (*Hemerocallis* varieties)
- *Epimedium grandiflorum*
- harlequin plant (*Houttuynia cordata* 'Chameleon')
- hostas (*Hosta* varieties)
- lungwort (*Pulmonaria* species and varieties, especially silver ones)
- sweet woodruff (*Galium odoratum*)

Shrubs

Shrubs have a lot to offer—flowers, fragrance, colourful berries and interesting bark and branches. Flowering shrubs offer the opportunity to have colour in our gardens virtually any time of year. Smaller gardens have forced us to limit the size of the plants we use, and gardeners want to spread the flowering times over a longer period. The downside of many flowering shrubs is their limited flowering period and rather plain appearance in the garden for the rest of the year. Well, hold onto your hat—breeders and growers around the world are continually producing new, more compact and better-looking flowering shrubs that are really appropriate for today's garden.

I like to arrange shrubs among flowering trees, colourful conifers and broadleaf evergreens like rhododendrons for a spectacular spring show. During the early part of winter, the branches of some flowering shrubs can be a source of cut flowers to enjoy indoors. Many shrubs, like the dwarf 'Golden Princess' spirea or the old-fashioned 'Goldflame' spirea, have great summer foliage colour. Shrubs can provide a sheltering thicket and some food for birds in winter, as well as attracting them all spring and summer to keep insects under control.

Calendar of flowering shrubs

Before you think about planting any flowering shrub, look at your calendar and pick out the months you need colour the most. Even though you may really enjoy certain plants, why have everything bloom in April and May, when with a little planning you can enjoy flowering shrubs year-round in zone 6 and warmer climates? Use the calendar to help choose shrubs for year-round interest.

January

Chinese witchhazel (*Hamamelis mollis*). Zone 6. Grows 2.4 to 4 m (8 to 13 feet). This shrub has very fragrant golden yellow flowers and yellow foliage in the autumn.

Wintersweet (*Chimonanthus praecox*). Zone 6. Grows 1.2 to 2 m (4 to 6½ feet). It has pale yellow flowers stained purple, with a strong fragrance, and blooms through February. It is best against a wall.

February

Cornelian cherry (*Cornus mas*). Zone 5. Grows 3 to 8 m (10 to 25 feet). It has small, yellow flowers, with no fragrance. In the fall, the cornelian cherry boasts brilliant colouring and bright, edible, cherrylike fruit.

February daphne (*Daphne mezereum*). Zone 3. Grows 1 to 1.2 m (3 to 4 feet). This shrub has small, rosy purple, very fragrant flowers along its stems. Its fruits are poisonous.

Spirea (*Spiraea thunbergii*). Zone 5. Grows to 60 cm (2 feet). It has small, white flowers with no fragrance. This is the earliest blooming spirea.

White forsythia (*Abeliophyllum distichum*). Zone 6. Grows 1 to 1.2 m (3 to 4 feet). This plant has very fragrant, tiny white flowers along its stems. It performs well in shady spots.

March

Burkwood viburnum (*Viburnum × burkwoodii*). Zone 5. Grows 1.8 to 3 m (6 to 10 feet). This shrub has pink buds and white, very fragrant flowers. It often blooms a second time in fall.

Flowering currant (*Ribes sanguineum*). Zone 6. Grows to 2.4 m (8 feet). The flowers of 'King Edward VII' are red; those of 'White Icicle' are white. The flowers are not fragrant, but do attract hummingbirds.

Flowering quince (*Chaenomeles japonica*). Zone 6. Grows 1.5 to 2.4 m (5 to 8 feet). It flowers red, pink, or white, and bears interesting applelike fruits in the fall.

***Forsythia × intermedia* 'Minigold'.** Zone 5. (See "Compact shrubs" listing.) Grows to 1.5 m (5 feet). It has yellow flowers.

Variegated forsythia (*Forsythia* 'Fiesta'). Zone 6. Grows 1.5 to 2 m (5 to 6½ feet). This plant bears yellow flowers with no fragrance along the long stems. It has variegated yellow and green foliage.

April

Garland spirea (*Spiraea × arguta*). Zone 3. Grows 1 to 2 m (3 to 6½ feet). White cascades of unscented flowers and blue-green foliage create an attractive contrast.

***Kerria japonica* 'Pleniflora'.** Zone 5. Grows to 2 m (6½ feet). The double yellow flowers have no fragrance. This shrub will sucker. Prune immediately after flowering.

Pearlbush (*Exochorda × macrantha*). Zone 5. Grows to 2 m (6½ feet). Large white, unscented flowers bloom along spectacular, long arching branches.

Redvein enkianthus (*Enkianthus campanulatus*). Zone 5. Grows 2 to 4 m (6½ to 13 feet). It boasts clusters of unscented, bell-shaped red flowers and brilliant fall colour.

May

Azaleas (deciduous varieties). Zone 5. Grows 1.5 to 2.4 m (5 to 8 feet). Most of these cultivars have brilliant fall colour. The Viscosums are very fragrant.

- 'Daybreak' has large, burnt orange flowers.
- 'George Reynolds' has large yellow flowers.
- 'Gibraltar' has large, brilliant orange-red flowers.
- 'Golden Eagle' has deep, bright orange flowers.
- 'Golden Sunset' has large, golden yellow flowers.
- 'Homebush' has rose-pink blossoms in globes.

- 'Klondyke' has large orange-gold flowers, tinted red.
- 'Oxydol' has large white flowers with a yellowish blotch.
- 'Satan' has deep geranium-red flowers.

Viscosum Hybrids
- 'Antilope' has pink and salmon flowers.
- 'Arpege' has deep yellow flowers.
- 'Rosata' has carmine-rose flowers.

Dwarf Korean lilac (*Syringa patula* 'Miss Kim'). (See "Compact shrubs" listing.) Zone 4. Grows to 1 m (3 feet). This is a very compact dwarf, great for small gardens. It has blue-purple, very fragrant flowers.

Fragrant snowball (*Viburnum* × *carlcephalum*). Zone 5. Grows 1.5 to 2 m (5 to 6½ feet). This shrub has pink buds, fragrant white flowers, and brilliant fall foliage.

French lilac (*Syringa vulgaris* varieties). Zone 3. Grows 3 to 6 m (10 to 20 feet). Flowers are very fragrant. Watch for leaf miners in early summer. Lime each fall in wet climates.
- 'Belle de Nancy' has double, pink flowers.
- 'Charles Joly' has double, dark red flowers.
- 'Katherine Havemeyer' has double, lavender flowers.
- 'Madame Lemoine' has double, pure white flowers.
- 'Souvenir de Louis Spaeth' has the best single, purple flowers.

Fringe flower (*Loropetalum chinense* 'Razzleberri'). (See "Compact shrubs" listing.) Zone 7. It has red, unscented flowers and deep burgundy foliage.

Maries' doublefile viburnum (*Viburnum plicatum* 'Mariesii'). Zone 5. Grows to 1 m (3 feet). Single, unscented lacecaps bloom in masses until frost. This shrub has red fruits and excellent fall colour.

June

Beautybush (*Kolkwitzia amabilis*). Zone 4. Grows 1.8 to 3 m (6 to 10 feet). The masses of pink, bell-shaped flowers have no scent.

Bridalwreath spirea (*Spiraea* × 'Van Houttei'). Zone 4. Grows 1 to 2 m (3 to 6½ feet). It has white flowers on arching branches.

Doublefile viburnum (*Viburnum plicata* 'Summer Snowflake'). Zone 6. Grows to 2 m (6½ feet). The white lacecap, unscented flowers last until frost.

Japanese spirea (*Spiraea japonica*). Zone 4. A popular choice.
- 'Anthony Waterer' grows to 60 cm (24 inches). Its cerise-pink flowers last from June until frost.
- 'Golden Princess' grows to 50 cm (20 inches). Its light pink flowers last until frost. This shrub also has beautiful gold foliage.
- 'Goldflame' grows to 50 cm (20 inches). It has cerise flowers and yellow foliage.
- 'Goldmound' grows to 50 cm (20 inches). It forms a beautiful gold mound and has light pink flowers that bloom until frost.
- 'Little Princess' grows to 50 cm (20 inches). It has light pink flowers that last until frost, green foliage, and a mounding habit.
- 'Magic Carpet' grows to 40 cm (16 inches). This very low shrub has golden coppery foliage and pink flowers that last until frost.
- 'Shirobana' grows to 60 cm (24 inches). Its bicolour pink and white flowers bloom until frost.

Tosa spirea (*S. nipponica* 'Snowmound'). Zone 4. Grows to 1 m (3 feet). It has many round clusters of white, unscented flowers.

Mock orange (*Philadelphus* hybrids).
- 'Virginal'. Zone 3. Grows 2.4 to 3 m (8 to 10 feet). It has fragrant, semidouble white flowers.
- 'Belle Etoile'. Zone 4. Grows up to 2 m (6½ feet). It has fragrant, single white flowers flushed maroon in the centres.

Potentilla, or shrubby cinquefoil (*Potentilla fruticosa*). The flowers are unscented.
- 'Pink Beauty'. Zone 3. Grows 60 to 80 cm (24 to 31 inches). The double, dark pink flowers hold their colour well.

- 'Red Ace'. Zone 3. Grows 60 to 80 cm (24 to 31 inches). Red in spring, the flowers turn orange in the heat of summer.
- 'Sutter's Gold'. Zone 3. Grows to 30 cm (12 inches). This shrub has bright yellow flowers and a low creeping habit.
- 'Yellow Gem'. Zone 4. Grows to 25 cm (10 inches). With bright yellow flowers, this is an outstanding groundcover.

Slender deutzia (*Deutzia gracilis*). Zone 5. Grows up to 1 m (3 feet). The white, unscented flowers cluster on arching branches.

Tamarisk (*Tamarix pentandra*). Zone 3. Grows 3.5 to 5 m (11 to 16 feet). Lacy pink flowers bloom on new growth. This is the most hardy tamarisk.

Weigela.
- Dwarf variegated weigela (*W. florida* 'Nana Variegata'). See "Compact shrubs" listing. Zone 4. Grows to 1 m (3 feet). Unscented pink flowers bloom again in late fall.
- *W. florida* 'Minuet'. Zone 4. Grows to 1.2 m (4 feet). This is a compact plant with bronze-tinged green leaves and unscented pink flowers.

July to October

Abelia 'Edward Goucher'. Zone 6. Grows 60 to 90 cm (2 to 3 feet). The unscented pink flowers last until frost.

Abelia × *grandiflora* 'Francis Mason'. (See "Compact shrubs" listing.) Zone 6. Grows 1 to 1.2 m (3 to 4 feet). The unscented pink flowers last until frost.

Butterfly bush (*Buddleia davidii*). (See "Old favourites" and "Compact shrubs" listings.) Zone 5. Grows to 3 m (10 feet). This is a continuing favourite for excellent reasons. The flowers are very fragrant and a variety of colours is available.

Henry's St. John's wort (*Hypericum patulum* var. *henryi*). (See "Compact shrubs" listing.) Zone 5. Grows to 1 m (3 feet). This shrub has golden flowers. It is evergreen in mild climates.

Hydrangea. (See "Old favourites" and "Compact shrubs" listings.) Zone 3 to 6. Grows .6 to 2 m (2 to 7 feet). Many varieties of this shrub are available. It's great for shady spots. Flowers are white, cream, pink or blue.

Rose-of-sharon (*Hibiscus syriacus*). Zone 6. Grows 1.8 to 3 m (6 to 10 feet). This shrub's single or double, unscented flowers bloom until September or October.

Shrubby St. John's wort (*Hypericum* 'Hidcote'). Zone 6. Grows to 2 m (6 feet). This semi-evergreen's yellow, saucer-shaped flowers with conspicuous red stamens bloom until October.

Sweetspire (*Itea virginica* 'Henry's Garnet'). (See "Best bets—compact and medium-sized flowering shrubs" listing later in chapter.)

November

Bodnant viburnum (*Viburnum* × *bodnantense* 'Dawn'). Zone 6. Grows 2.4 to 3 m (8 to 10 feet). Fragrant, pink flowers bloom in clusters from November to April.

Harry Lauder's walking stick (*Corylus avellana* 'Contorta'). Zone 4. Grows 2.4 to 3 m (8 to 10 feet). This shrub's attractive pendulous yellow catkins grow longer as winter progresses.

December

Chinese witchhazel (*Hamamelis mollis*). See January.

Old favourites—the backbone of the garden

Broom. Zone 5. The tall flowering broom is often used the wrong way in gardens. The popular soft yellow *Cytisus praecox* and the red-flowering *C. scoparius* 'Burkwoodii' need a very hard shearing after they finish flowering. Broom makes a great groundcover if kept pruned in low, oval mounds. There are dwarf varieties, such as the yellow *C.* × *kewensis* and *C. purgans*.

Buddleia. Perhaps the most fragrant of the summer-blooming shrubs is *Buddleia davidii*. It has been dubbed the butterfly bush for good reason; on sunny days, you will seldom see this plant without a butterfly nestled on one of its blossoms. Starting in July, long flower spikes will fill your garden with perfume. The standard varieties grow up to 3 m (10 feet) with white, blue, deep purple, pink and soft yellow flowers. They should be cut back to about ground level each spring to keep them low and bushy. Lower-growing varieties were introduced a few years ago, and you can enjoy them in tighter spaces as well. (See "Best bets—compact and medium-sized flowering shrubs" below.)

Deciduous azaleas. The old Mollis varieties still pack a wallop of colour, with their brilliant oranges and reds. If you can track down one of the yellow seedling varieties, your garden will be filled with a sweet perfume. Speaking of fragrance, you might also try the Viscosum Hybrids. They are sometimes called swamp honeysuckles because of their fragrance and narrow, funnel-shaped, smaller flowers. For some special varieties to try, see the "Calender of Flowering Shrubs" earlier in this chapter.

Hibiscus. Hardy garden hibiscus or rose-of-sharon (*Hibiscus syriacus*) is every bit as beautiful as its tropical counterparts and yet hardy to zone 6. Growing anywhere from 2 to 3 m (7 to 10 feet) high, rose-of-sharon will flower from July until October. Try the double varieties, like magenta-rose 'Collie Mullens' or soft pink and red-throated 'Blushing Bride'. I'm really fond of the singles, such as 'Red Heart', a pure white with a scarlet centre, 'Blue Bird' and magenta-rose 'Woodbridge'. Most varieties have a deep-coloured centre that gives the flower a bicolour effect.

Hydrangea. Hydrangeas are popular because of the fresh new colour they bring to a summer and early fall garden. They also do well in those difficult shady locations. Although generally considered shade plants, hydrangeas will do well in full sun, but tend to be more compact and bloom earlier, especially in periods of extreme heat. Most varieties need lots of moisture to look lush and healthy through the growing season.

Hardy varieties:
- Hills-of-snow (*H. arborescens* 'Annabelle'). Zone 3. Grows to 2 m (7 feet). Small shrub with loose, bushy growth and creamy white, round flowers from July to September.
- *H. paniculata* 'Kyushu'. Zone 4. Grows to 2 m (7 feet). An elegant hydrangea, with deep cream, 20-cm (8-inch), cone-shaped, almost lacy blossoms. Blooms July to September and has dark, glossy leaves.
- Peegee hydrangea (*H. paniculata* 'Grandiflora'). Zone 4. Grows to 2 m (7 feet). One of the showiest white-flowering autumn shrubs. Massive, cone-shaped blossoms bloom July to October, turning pink when mature.

Lacecap varieties are similar to mopheads in their growth habits and soil requirements, but the flowerheads have a delightful lacy effect. The centre flowers are sterile buds surrounded by rows of large single florets. Zone 6.
- 'Blue Wave' is one of the best lacecaps that can change colour with the pH. Plant in semishade.
- 'Lemon Wave' has colourful, deep green leaves with distinct bright yellow markings. White lacecap flowers turn pink as they mature. Best in coastal zones.

Mophead varieties (*Hydrangea macrophylla*) average 1.2 to 1.8 m (4 to 6 feet) in height and are rated zone 6. Colours can be changed by adjusting the pH of the soil. For pink, add lime. For blue, add aluminum sulphate.

- 'Forever Pink'. Compact pink mophead with a more continuous blooming habit.
- 'Nikko Blue'. One of the best mopheads, turning purple in acid soils.
- 'Soeur Thérèse'. A pure white mophead that does best in light shade.

For the very compact hydrangeas, see "Best bets—compact and medium-sized flowering shrubs" below.

Viburnum. From the old-fashioned snowball tree (*Viburnum opulus* 'Sterile') to double-file viburnum (*V. plicata* 'Summer Snowflake'), which blooms from May or June until frost, viburnums are a highlight of the spring garden. If you enjoy fragrance, fragrant snowball (*V.* × *carlcephalum*) will knock your socks off. I'm also impressed with the semi-evergreen Burkwood viburnum (*V.* × *burkwoodii*), with its glossy leaves and fragrant, pink-budded blossom clusters, which open into pure white, perfumed balls.

Best bets—compact and medium-sized flowering shrubs

Dwarf butterfly bush (*Buddleia nanhoensis* 'Petite Indigo' and 'Petite Plum'). Zone 5. Both have grey foliage, but 'Petite Indigo' has fragrant lilac-blue flowers, while 'Petite Plum' has reddish flowers that bloom July through September. Both attract butterflies.

Dwarf deutzia (*Deutzia gracilis* 'Nikko'). Zone 6. Extremely dwarf; grows to 50 cm (20 inches) high and spreads to about 1.5 m (5 feet). Green leaves turn deep burgundy in fall. Double white blossoms in May.

Dwarf Korean lilac (*Syringa patula* 'Miss Kim'). Zone 4. Slow growing to 1 m (3 feet) high by 1 m (3 feet) wide. Small, dark green leaves with a bronze tint. Small, fragrant blue-purple flowers in June.

Dwarf mock orange (*Philadelphus virginalis* 'Dwarf Minnesota Snowflake'). Zone 4. Grows 1 m (3 feet) high and is smothered each June in fragrant, double white blooms.

Dwarf spireas (*Spiraea* varieties). Zone 4. My favourites, growing 40 to 90 cm (16 inches to 3 feet) high and mounding, all bloom with pink flowers June through to frost. All are a must for any landscape—as an underplanting or groundcover. 'Limemound' has chartreuse-yellow foliage; 'Golden Princess' has golden yellow foliage; 'Little Princess' has green foliage; 'Magic Carpet' has golden bronze foliage. (For more varieties, see the "Calendar of Flowering Shrubs" above.)

Dwarf variegated weigela (*Weigela florida* 'Nana Variegata'). Zone 4. One of the best variegated deciduous shrubs you'll find. Growing only 1 m (3 feet) high and about the same in width, it produces soft pink blossoms in late spring and sporadically in late fall.

Forsythia (*Forsythia* × *intermedia* 'Minigold'). Zone 5. Very compact to 1.5 m (5 feet) and produces yellow flowers in February and March. Prefers full sun to partial shade.

Fringe flower (*Loropetalum chinense* 'Razzleberri'). Zone 7. This Chinese introduction is magnificent year-round. Its deep burgundy foliage and red, fringelike flowers appear spring through fall.

Henry's St. John's wort (*Hypericum patulum* var. *henryi*). Zone 5. A seldom-used plant that produces bright golden flowers from summer to fall. Loves full sun, grows only 1 m (3 feet) high and is evergreen in mild climates.

Japanese flowering quince (*Chaenomeles japonica*). Zone 5. 'Cameo' has soft apricot-pink double flowers in March and only grows to about a metre (3 feet). It produces applelike fruits in fall. They taste awful but make wonderful jelly.

Miniature French hydrangea (*Hydrangea macrophylla* 'Pink Elf'). Zone 6. 'Pia' is one of the very best dwarf hydrangeas, growing only to 60 cm (24 inches). Bears small, profusely produced pink to mauve mophead flowers. Loves partial shade and performs well in containers.

Shrubby cinquefoil (*Potentilla fruticosa*). Zone 3. One of the most underrated flowering shrubs in coastal areas. They flower late May through October. Lower-growing varieties are the best for most gardens. 'Yellow Gem', a University of British Columbia introduction, is low and compact. 'Red Robin' is compact, with orange-red blossoms, and 'Pink Beauty' has semidouble pink blossoms that hold their colour through the summer.

Sweetspire (*Itea virginica* 'Henry's Garnet'). Zone 5. A virtually unknown plant with slightly perfumed, white flower spikes in summer. Grows 1 to 1.5 m (3 to 5 feet) with green foliage that turns red-purple in fall.

Variegated glossy abelia (*Abelia* × *grandiflora* 'Francis Mason'). Zone 6. Beautiful yellow and green variegated foliage with bronze leaves and soft pink flowers all summer. Grows 1 to 1.2 m (3 to 4 feet) high.

Best bets—shrubs for berries

In winter, when most deciduous plants are at their low point, colourful berries provide a lift in the landscape and feed overwintering birds. Not all berries last through the winter; some disappear after severe frost and others have fallen by late December.

Beautyberry (*Callicarpa bodinieri* 'Profusion'). Zone 6. Purple berries on 3-m (10-foot) plants attract birds and are useful for holiday decorating.

Shrub roses (*Rosa rugosa* hybrids). Zone 3. Orange hips are great for making tea, for use in jellies, and for attracting birds. Grow to about 2 m (7 feet). (See chapter 7 for details.)

Snowberry (*Symphoricarpos albus*). Zone 4. White or pink berries last all winter long if they're not eaten by the birds. They're attractive with holly.

Winterberry (*Ilex verticillata*). Zone 6. Red berries on 3-m (10 foot) shrubs. Male and female plants are needed for berries. They attract birds and the intensely red berries are a natural for Christmas decorating.

Best bets—shrubs for colourful branches

Bloodtwig dogwood (*Cornus sanguinea* 'Midwinter Fire'). Zone 5. Red and yellow twigs and branches. Grows 1.5 to 2 m (5 to 7 feet).

Red twig dogwood (*Cornus atrosanguinea* 'Sibirica'). Zone 2. Red stems are particularly attractive in winter. Grows 1.5 to 2 m (5 to 7 feet).

Yellowtwig dogwood (*Cornus stolonifera* 'Flaviramea'). Zone 2. Yellow stems and twigs. Grows 1.5 to 2 m (5 to 7 feet).

Problems and Solutions

Mixing trees and pavement

It is possible to have a patio close to quite a large tree, but there are a few things you can do that will mitigate damage to the tree. Use paving stones rather than a solid slab of concrete. The stones are set in sand and will allow water to flow through to the tree roots.

If you are going to raise the soil around the base of a tree, leave an area surrounding the tree at its original level. This area (called a well) should be 1 to 1.8 m (3 to 6 feet) in diameter. You can create it by using landscape ties, stones or bricks. Most gardening or building centres have a number of options to choose from.

Put an iron grate around the base of the tree. It gives a finished look to the area and allows water to get through. Rather than raising the soil level and planting flowers at the bottom of trees, install an iron grate and place flowers in containers at the base.

Straightening a leaning tree

If you have a fairly young tree that's been pushed over by the weight of snow or some other circumstance, you can straighten it out. Pound a strong pole or galvanized pipe deep into the ground opposite to the direction it is leaning. Then tie the tree to the pole, using a broad strap that won't cut into the bark, such as an old garden hose.

On larger trees, you should first dig a trench around the tree as if you were going to dig it up, making a clean cut to prune the roots. This will allow new fibrous roots to develop, while the pole holds it securely.

Preventing snow and ice buildup on trees

In some situations, heavy snow and ice can damage trees. The weight can break branches or even split the tree. If you're worried, knock the snow off from underneath with a broom. Be aware, though, that once trees are frozen solid they become very brittle and you can do more harm than good.

As a preventive measure, you can help support the branches by tying them up with

Troubleshooting tree and shrub problems

When a tree or shrub shows signs of distress it can be tough to figure out the problem right away. It could be caused by weather conditions, pests or some sort of disease. Here's how to be a good tree detective.

Think about the tree's environment.

- Has anything changed lately in its surroundings? Is there a new water line that may have interfered with its roots, for example?
- Is it standing in a pool of water that wasn't there before?
- Has the neighbouring vegetation changed?
- Have you given it different nutrients lately?
- Have any herbicides been sprayed in the area?
- Has the weather been unusual? Hot? Cold? Sudden temperature changes?

Any of these changes could be related to the problem.

Get right in there and have a good look at the tree. Inspect it from the ground up. Look carefully at the bark, stems and all sides of the leaves or needles for clues. Where does the problem occur—leaves, stems, bark or all three?

- Read the leaves. Are there deposits (hard, sticky, gooey) on the leaves? What about tiny bugs or wispy caterpillar nests? If there is discolouration, note its shape, colour, and where it is.
- Sound out the bark. Scrape the bark and see if there is any life in the wood. There should be green in the cambium layer, just under the bark. Is the bark splitting or peeling?
- Root around. Dig down and look at the roots. New root growth should be apparent and it should be white and healthy looking. If the roots are turning brown and mushy, this is an indication of root rot. If the tree has root rot, there's nothing you can do, unless it's small enough to move to a better-drained location.

DECIDUOUS TREES AND SHRUBS

something like binder twine. Run it around the outside of the branches to hold the plant more tightly together.

Assessing winter damage

Trees can be damaged by a sudden, drastic turn in the weather. This is especially true in the spring, when a mild spell can get the sap flowing. If this is followed by a sudden blast of cold, you may have a problem.

Closer to spring, especially after a bit of warm weather, cut the dead wood back to where there is life. I always go to the tips of the plant on the outside branches and look to see if there is new growth coming. If there are buds, all will be well. In extreme cases, some plants will freeze right to the ground and you'll see nothing but dead branches and burnt leaves. But all is not lost yet—sometimes these frost victims will come back from the roots. If July rolls around and you're still waiting, it's time to give up!

Removing moss from trees

Moss is a natural inhabitant of our woodland gardens. However, moss on a tree can damage the cambium layer just below the bark and weaken the tree. You can brush it off with a scrub brush, or spray it with a lime sulphur spray. Avoid using lime sulphur on maples.

Spraying to prevent pest and disease problems

I often recommend the use of a dormant spray in winter. It is primarily organic and composed of lime sulphur and dormant oil. Sulphur kills spores of overwintering fungal diseases, as well as moss and lichen. The oil controls insect eggs and larvae that have been laid in the trees by sealing them off and cutting off the air supply. This destroys them, and the insects won't hatch in spring.

Spray three times, ideally a month apart, in November, December and January. Winter weather often dictates otherwise, so the key is

Warning

Sulphur, oil and copper sprays can stain plants and buildings. Cover anything you don't want sprayed with plastic covers to prevent staining or damage.

I recommend using warm water and a pressure sprayer to penetrate the tree's bark. The second-best option is a hose-end sprayer that will accommodate the concentrated oil and sulphur.

Always spray to run-off, which means you apply the product until the spray begins to run off the branch. It's important that the spray is not washed off by rain or frozen in place before it has a chance to work—that's why you really need to check with the weather office before you get out the sprayer.

Whenever you spray you should wear a proper mask, as well as rubber gloves and suitable clothing, such as coveralls—especially if you have allergies or chemical sensitivities. Before releasing the pump on pressure sprays, be sure the nozzle is pointed away from you.

to have three sprayings, at least 10 days apart. It should not be windy, freezing or raining when you do it—a tall order in winter, but it can be done.

You can use dormant sprays during the summer, but the rules change. Spray the oil and the lime sulphur separately and cut the strength according to package directions for growing season applications.

Note: The oil cannot be used on all trees. Maples, nut trees (walnuts in particular) and viburnums do not like this spray.

Copper spray is an alternative to the lime sulphur and oil treatment. It works better in controlling peach leaf curl on peaches and nectarines, and helps with the European canker on apple trees. European canker causes the bark to split, eating into the cambium layer and eventually killing the branches.

Correcting fertilizer burn

Brown tips on new growth usually indicates a burn caused by too much fertilizer. The solution is to leach the excess fertilizer out of the soil by saturating it to the drip line of the tree. Leave a soaker hose on overnight.

Healing tree wounds

Sometimes limbs that are pruned fail to heal and continue to drip sap. This is usually the result of pruning at the wrong time of year—especially when the sap is flowing. I don't recommend painting the exposed area: tests have shown little benefit from this technique. Instead, try covering the cut area with black plastic. The darkness helps the tree to heal and eventually stop the flow.

Planting under trees

Many folks want lawn grasses to come right up to the trunks of trees. The moisture pulled out of the soil by larger trees (up to 450 litres/ 100 gallons every day), and the competition with the tree roots sometimes makes growing lawn grasses next to impossible. This is where groundcovers play such an important role.

This may sound weird, but I recommend planting variegated bishop's weed (*Aegopodium podagraria* 'Variegatum') at the base of trees. It will give any tree—or weed, for that matter—a run for its money. The green and white foliage is attractive until it dies back in late October, and it will come back again in the spring. Here are a few other good bets for tough groundcovers.

- Japanese spurge (*Pachysandra terminalis*). Zone 3.
- Kinnikinnick (*Arctostaphylos uva-ursi*). Zone 4. Evergreen, native to B.C.
- Periwinkle (*Vinca minor*). Zone 6. Evergreen.
- Baltic ivy (*Hedera helix* 'Baltica'). Zone 5. The hardiest of ivies.

Cutting down a tree

Eliminating a tree from your landscape often goes beyond cutting the tree down. That's because the roots try to regenerate by sending up new growth. There are a number of ways to tackle this, depending on the size of the problem. You can cut and remove the suckers as they appear, but the roots can send up new suckers for years afterwards. Here's how to keep ahead of the problem.

- Spot spray the leaves of the suckers with a herbicide, which will work its way down to the root system and eventually kill the roots.
- Remove the stump, if possible.
- Drill a hole in the stump and use a stump remover product, which speeds up the process of rotting.
- Root prune the remaining stump so the roots will not send up sucker growth. This is a good solution and not as onerous a task as removing the stump. To root prune, dig a trench in a circle around the tree, cutting all root growth as you go. Use a sharp spade, a saw or shears, depending on root size.

Broadleaf Evergreens and Conifers

This is without a doubt the most valuable plant grouping for any moderate climate, zone 6 and above. Broadleaf evergreens are plants that have green foliage all year, but are not evergreen conifers. They encompass a wide range of plants—including heathers, azaleas, rhododendrons, bamboo and holly. Conifers are an essential component of any garden, as feature trees or hedging.

Broadleaf Evergreens

Broadleaf plants offer the diversity of year-round foliage that changes colour with the seasons and flowers that appear seasonally. Many broadleafs, including skimmia, pyracantha, and pernettya, have attractive berries that last through the winter to spark up those dreary short days.

Rhododendrons and azaleas are certainly the best-known broadleaf plants, but there are many others that provide flowers, berries or colourful foliage every month of the year, so there's always something to look forward to in your garden.

Unfortunately, there are very few broadleaf plants that will tolerate the bitterly cold winters of zone 4 and below. Rock daphne (*Daphne cneorum*), with its fragrant pink blossoms, is one exception.

Calendar of broadleaf shrubs

Zone 6 and above is where broadleafs can tolerate most winters and still provide a valuable year-round show. In protected areas of zone 5, you can squeeze in a few more, such as cotoneasters, hardy rhododendrons and azaleas, and hardy hollies. Rhododendrons and azaleas are not included in this listing because they have their own section later in this chapter.

Use the calendar of shrubs below to plan flower and berry interest all year round in your garden.

January

Erica × darleyensis **'Silberschmetze'**. Zone 8. White flowers with a light scent bloom October to April.

Winter-flowering heather (*Erica carnea*). Zone 6. Grows to 15 to 30 cm (6 to 12 inches).

- 'Golden Starlet' has gold foliage and white flowers, and blooms November to April.
- 'Kramer's Red' has the deepest rose colour of any heather and flowers from October to April.
- 'Myretoun Ruby' has deep rose flowers in bloom from October to April.

February

Himalayan sweet box (*Sarcococca* species). Loves shade.

- *S. hookerana* var. *humilis*. Zone 5. A wonderful groundcover that usually grows 40 to 60 cm (16 to 24 inches). Small, shiny leaves and tiny white, intensely fragrant flowers. Produces black berries.
- *S. ruscifolia*. Zone 5. Grows to 1.2 m (4 feet). Has red berries, and tiny white flowers emit a sweet scent in February. It's a good foundation plant if pruned.

Silk tassel (*Garrya elliptica*). Zone 7. Bears long pendulous catkins February to April. Grows 2.4 to 3 m (8 to 10 feet).

Winter rhododendron (*Rhododendron × praecox*). Zone 6. Bears small mauve flowers February to March. Grows 1.2 m (4 feet). Plant it where it is protected from cold winter winds.

March

Lily-of-the-valley bush (*Pieris japonica*). Zone 6. My all-time favourite broadleaf. There's not a time of year when it doesn't look great and it tolerates sun or shade. Prune hard in the growing season.

- 'Cavatine'. Zone 5. Growing to 60 cm (2 feet), this is one of the most outstanding varieties. Blossoms often cover the entire plant. As a bonus, it does not set seed to distract from the plant.
- 'Flamingo' has long clusters of deep pink flowers, bronze summer colour and grows 1.5 to 2 m (5 to 7 feet).
- 'Havilla' tolerates deep shade and variegated white and green leaves add vitality to shady spots. New growth is a wonderful red. This is an outstanding plant. Grows 70 to 90 cm (28 to 36 inches).
- 'Mountain Fire' is the toughest variety. White flowers, brilliant red growth in spring through summer. Grows 1.5 to 2 m (5 to 7 feet).
- 'Prelude' is a very compact form, 1.5 to 2 m (5 to 7 feet). Blooms very late; a profusion of short stalks bear pure white flowers. New growth is bright pink.

roadleaf evergreens for berries

Berry plants are a treat in the winter months when we really need a colour boost in our gardens. They provide colour, serve as screens, groundcovers and feature plants, and many of them are a great source of natural food for overwintering birds. Broadleafs such as holly and skimmia provide wonderful Christmas greens. Some, like wintergreen, are delicious as a refreshing midwinter snack. Berries on broadleaf plants are perhaps the most stunning because the evergreen leaves, often in rich winter colouring, provide a nice accent. (See the "Calendar of broadleaf shrubs" and "Broadleaf evergreens for a shady spot" for details on the shrubs below, all of which have great berries.)

- Chilean pernettya
- cotoneaster
- firethorn
- holly
- Oregon grape
- *Skimmia japonica*
- strawberry tree
- wintergreen

Oregon grape (*Mahonia* varieties). Zone 5. These plants, with yellow flowers early in the year, colourful winter foliage and berries, are a must! They all do well in the shade.
- *M. aquifolium* grows to 1.8 to 2.4 m (6 to 8 feet). It bears spikes of yellow flowers and later produces tiny blue berries.
- *M. nervosa* is lower growing, with less glossy leaves.
- *M. repens,* at 15 to 80 cm (6 to 30 inches) makes a good creeping groundcover.
- *M.* × *media* 'Charity' (zone 7) is very tall growing and bears attractive, erect spikes of fragrant, deep yellow flowers from late December to spring.

Wintergreen (*Gaultheria procumbens*). Zone 4. A shade-loving groundcover to 15 cm (6 inches). Likes acid soil and is a good companion to lily-of-the-valley shrub and rhododendrons. Needs fine bark mulch in the soil to do its best. Small white flowers in spring are followed by delicious, plump red berries. Leaves and fruit are aromatic when crushed.

April

Daphne × *burkwoodii* **'Carol Mackie'.** Zone 4. Light pink, very fragrant flowers appear in late April or early May on outstanding variegated cream and green foliage. Needs good drainage and grows 1 to 1.5 m (3 to 5 feet). 'Moonlight' is a zone 6 variety with outstanding, vibrant foliage colour—bright cream-yellow with green leaf edgings (the reverse of 'Carol Mackie'). Fragrant flowers in May, grows 1 to 1.5 m (3 to 5 feet).

Rock daphne (*Daphne cneorum*). Zone 2. 'Ruby Glow' is one of the most intensely fragrant plants for the garden, with masses of deep pink, clustered flowers. It will tolerate some shade. Must be pruned hard after blooming to maintain shape. Grows .5 to 1 m (20 to 40 inches).

May

California lilac (*Ceanothus thyrsiflorus*). 'Victoria' is one of its sturdiest hybrids and will survive in a protected zone 6 climate. It grows to a relatively large 3-m (10-foot) size, but can easily be kept compact with pruning. Masses of blue flower spikes appear in May and will continue straight through until September.

Creeping broom (*Genista pilosa* 'Vancouver Gold'). Zone 5. A fabulous low shrub or groundcover with masses of tiny, pealike blooms May into June. Attractive all year, it grows 10 cm (4 inches) high by about 1 m (3 feet) wide.

June

Mountain laurel (*Kalmia latifolia*). Zone 4. One of the outstanding flowering plants in June. Plant it near rhododendrons to extend the display. Needs excellent drainage and does well in poor soil. Buds have intense colours, opening to saucerlike blossoms. Most grow 1.2 to 1.5 m (4 to 5 feet) and tolerate shade or sun.
- 'Little Linda' is a miniature that grows to only 90 cm (3 feet). Red buds open to pink. Has glossy foliage and a round shape.
- 'Nipmuck' has intense red buds that open to soft pink.
- 'Olympic Wedding' has pink buds, opening to a burgundy band on a pink corolla.
- 'Ostbo Red', one of the first named cultivars, has deep red buds, opening to deep pink flowers.
- 'Richard Jaynes' has deep pink buds, opening to rich pink.
- 'Snowdrift' has white buds and pure white blossoms.

July

Adam's needle (*Yucca filamentosa*). Zone 4. These plants make a wonderful focal point in any garden and the new variegated forms are breathtaking. Tall spikes of bell-shaped flowers appear in July and August. They need good drainage and do best if protected from cold winter winds.

- 'Golden Sword' has striking foliage—yellow centres with a soft green margin—and fragrant white summer flowers. Grows 60 to 90 cm (2 to 3 feet) tall.

August
Himalayan honeysuckle (*Leycesteria formosa*). Zone 6. This shrub has multiple hollow green stems that produce masses of drooping flower clusters, the wine-coloured bracts contrasting with the white petals. These 1 to 2-m (3- to 7-foot) plants start blooming in summer and are prolific in August and September.

Scots heather (*Calluna vulgaris*). Zone 6. Most grow 30 to 90 cm (1 to 3 feet) and bloom late summer to fall.

- 'Alba Plena'. Grows 30 to 45 cm (12 to 18 inches). The best double white.
- 'County Wicklow'. Grows to 25 cm (10 inches). Double, shell-pink flowers.
- 'Peter Sparkes'. Grows 60 to 86 cm (24 to 34 inches). Double, deep pink flowers are good for cutting.
- 'Tib'. Grows 30 to 45 cm (12 to 18 inches). Double, dark red flowers.

September
Chilean pernettya (*Pernettya mucronata*, now called *Gaultheria mucronata*). Zone 7. One of the most magnificent berried plants. Female, berry-bearing plants must be polli-

Broadleaf evergreens for a shady spot

The flowers, shiny foliage, variegation and berries of many broadleaf plants do much to lift and brighten those dark, dull areas.

- *Euonymus fortunei* 'Gaiety'. The old reliable, with green and white, uplifting foliage.
- *Leucothöe fontanesiana* 'Rainbow'. Variegated foliage is a treat.
- Lily-of-the-valley shrub (*Pieris japonica*). Great foliage; the variegated forms tolerate heavy shade. (See the calendar—March.)
- Mountain laurel (*Kalmia latifolia*). Intense red and pink buds make a great show in June and July. (See the calendar—June.)
- Oregon grape (*Mahonia aquifolium*, *M.* × *media* 'Charity' and *M. repens*). Early flowers and colourful winter foliage—a must! (See the calendar—March.)
- *Prunus laurocerasus* 'Otto Luyken'. A great foundation plant with white, fragrant flowers and handsome appearance.
- *Skimmia japonica,* male and female. Bronze buds and female berries match up to create a very nice couple, indeed. (See the calendar—November.)
- Sweet box (*Sarcococca hookeriana* var. *humilis*). Delightful fragrance in February. (See the calendar—February.)
- Variegated Algerian ivy (*Hedera algeriensis* varieties). Huge yellow-blotched leaves are spectacular.
- *Viburnum davidii*. Bronze buds, white flowers and blue berries. A winner.

nated by male. Pinkish, bell-shaped flowers appear in late spring. Huge white, pink or red berries smother the plant in September and continue through winter. Needs a sunny, well-drained location and winter protection. Grows 60 to 90 cm (2 to 3 feet).

Cotoneaster (*Cotoneaster* varieties). Zone 5. These plants bear small white flowers in May and June but it is the fall display of bright orange to orange-red berries that is really eye-catching.

- *C. dammeri* is a good groundcover that grows to 10 cm (4 inches) by 1 m (3 feet). Its cultivar 'Eichholz' grows to 20 cm (8 inches) by 1 m (3 feet).
- *C. franchetii* (zone 6) is a tall-growing plant, to 2 m (7 feet), with silver-blue foliage.

October

Firethorn (*Pyracantha* hybrids). Zone 6. These thorny, fast-growing plants are best espaliered against a wall or trained up a pillar. Berries are yellow, orange or red. If you're in a wet climate, be sure to select a variety that resists mildew and other fungal diseases. Massive white flowers appear in May. It's a great security plant thanks to its thorns, and it needs annual pruning.

- 'Orange Glow' is a favourite for its hardiness and disease tolerance. Grows 2.4 to 3 m (8 to 10 feet).
- 'Rutgers' is a prostrate variety that has orange-red fruit and is blight resistant. Grows to 1.2 m (4 feet).
- 'Teton' is a very upright and compact variety, to 2 m (7 feet), with yellow-orange berries. It requires less pruning than some of the others.
- 'Golden Charm' is one of the better yellow-berried varieties. Grows to 2.4 m (8 feet).

Strawberry tree (*Arbutus unedo*). Zone 7. Fabulous, large, strawberry-red berries adorn this slow-growing shrub in the fall. Eventually gets to 2.4 to 3.6 m (8 to 12 feet). Fruit and flowers are borne at the same time. Leaves are dark and glossy. Likes full sun.

November

Skimmia japonica. Zone 7. The more I see these plants in the landscape, the more I appreciate them. They love full shade to partial sun and provide slightly perfumed rose-red flowers in spring and attractive red berries in fall and winter. Unless you have a self-fertile variety, such as *S. japonica* ssp. *reevesiana,* 60 to 90 cm (2 to 3 feet), you need both male and female plants for pollination (one male will take care of up to nine females in this plant harem). There are both compact and standard varieties and they can make a big difference in your landscape.

- 'Rubinetta' is a wonderful male clone with intense red buds all winter. Grows to 60 to 90 cm (2 to 3 feet).

December

Holly (*Ilex* varieties). Zones 4 to 7. Perhaps one of the most underutilized plants in the landscape. Most of the new varieties are self-fertile, producing flowers followed by berries in late September that last well into the spring. I prefer to plant the variegated forms for foliage as well as berry colour. Most English holly (*Ilex aquifolium*) varieties grow 3 to 5 m (10 to 17 feet).

- *Ilex aquifolium* 'Siberia' (zone 5) is one of the hardiest self-fertile varieties. *I. aquifolium* 'Argentea Marginata' and 'Aurea Marginata' are female variegated forms that are attractive, but need a male for pollination. 'San Gabriel', 'Ebony Magic' and 'Sparkler' are self fertile. They grow 3 to 5 m (10 to 17 feet).
- The Blue Holly Group includes *Ilex* x *meserveae* 'Blue Boy' and 'Blue Girl'. They need pollinators but are hardier (zone 4), have unique blue foliage and only grow 2 to 2.7 m (7 to 9 feet). This makes them

ideal shrubs and hedges for small-space gardens. Plant males and females together to conserve space.

- 'J. C. Van Tol' is a superb cultivar with shiny, smooth, spineless leaves and large berries. It does not need a male pollinator. The smooth leaves make it easy to work with.

Evergreen azaleas

Evergreen Japanese azaleas are some of the most valuable broadleaf plants in any garden, especially in small spaces. They have a wonderful natural shape, flower spectacularly and some have attractive bronze leaves in winter.

Evergreen azaleas are great by themselves, especially when planted in groupings of three or more. They also combine well with other plants, such as the dwarf evergreen perennial candytuft (*Iberis sempervirens* 'Purity' or 'Little Gem') and winter-flowering heather. Planted under Japanese maples or flowering cherries, they make a great groundcover.

For optimum results in moist climates, azaleas require an open, porous, well-drained soil. Contrary to popular belief, they love full, hot sun as long as they are not up against a reflective wall on the hot south or west sides of buildings. Partial shade during the hottest part of the day is ideal, but they do not perform well in heavy shade.

There are more than 10 main groups of evergreen azaleas, which vary in blossom size, hardiness and overall size. My personal favourites are the Kurume azaleas because of their compact growing habit, tiny leaves, massive small-flower display and hardiness. Most evergreen azaleas will tolerate zone 6 and range in height anywhere from .6 to 1.2 m (2 to 4 feet).

Azaleas that get a little too large for their garden location can be sheared heavily, immediately after flowering, just as the new growth begins. Pruning hard old wood is not a problem, as they send new growth out of it. Bud set takes place in late August through October, so late-summer pruning, except to even out a few leggy branches, is not a good idea. A quality rhododendron and azalea fertilizer is all they require early in the spring before blooming, and again in late summer for bud set.

Midseason = mid- to late April   Late = May and later

Name	Season	Flower	Comments
'Blue Danube'	mid	intense blue violet	Single flowers.
'Girard's Fuchsia'	mid	reddish purple	Single flowers with wavy petals.
'Herbert'	mid	lavender	Orchidlike double flowers.
'Hino Crimson'	mid	crimson	Vibrant, single flowers. Plant with dwarf variety of white iberis.
'Hino-White'	mid	white	Fabulous display of tiny blossoms. Very compact.
'Purple Splendor'	mid	deep purple	Semidouble.
'Roberta'	mid	pink	Large, 8-cm (3-inch) double blossoms with ruffled petals.

Name	Season	Flower	Comments
'Macrantha'	late	salmon	Single blossoms. Low, spreading plant. Blooms after most other varieties have finished, often into June.
'Red Fountain'	late	dark red-orange	It's the last azalea to bloom, often into early July. Has a low, creeping form and will trail over rocks and walls.
'Rosebud'	late	rose-pink	Double blossoms with a rosebudlike appearance. Low, spreading form.

Rhododendrons

Most folks think of rhododendrons as those wonderful, big plants that bloom in April and May with a wide range of colours—from pink and lavender to brilliant red, pure white and deep purple. But there's a whole lot more! The genus *Rhododendron* has about 800 species, ranging from prostrate alpines to giant trees with huge leaves. In addition, there are over 10,000 named varieties—a number that increases yearly as new hybrids are developed.

All of these hybrids and species make it very confusing for the home gardener, so I have organized the listings here according to size. There are three categories: very compact and low growing, generally under 1 m (3 feet); medium, 1 to 2 m (3 to 7 feet); and the big guys, topping the charts at 2 m (7 feet) and over. These sizes are based on 10 years of growth. The final size of the plant should be at the top of your list when you're choosing a variety. Take a good look and decide how big a rhodo you want a few years down the road. Once that's looked after, you can get on to other considerations, such as flowers, leaf form and fragrance.

On a trip to England a few years back I was reminded of the spectacular beauty of the larger varieties. Some were as tall as double-decker buses and many had been growing for more than a century. Even though they weren't in bloom at the time, I was awestruck by these massive specimens, and could only imagine what they would look like when their masses of buds burst into bloom in the spring. These giants don't always fit into the home garden, though, and that's why smaller-growing varieties are so popular today.

A range of sizes and colours permits great versatility in the landscape, and I can only encourage you to take advantage of the opportunities presented by rhododendrons. Rockeries, borders and groupings of other plants are all suitable locations for rhodos. The dwarf varieties lend themselves to low-maintenance situations. The only obstacle you might have to overcome is paying the price—a dwarf, 15-cm (6-inch) plant costs the grower the same amount to produce as a 1-m (3-foot) plant of a different variety. In other words, size doesn't always represent the amount of work put into producing a plant.

Fragrant foliage is an attribute shared by many of the tiny-leafed varieties of rhododendrons, but 'Purple Gem' is one of the best, with deep purple flowers and new foliage that takes on a blue shade. It's hardy to −30°C (−25°F), which allows it to be used in cooler climate zones. It grows to about 60 cm (24 inches).

Rhododendrons require acidic soils if at all possible, but more importantly, they require open, porous soil to allow air into the root zone. This porosity becomes even more important with the other major requirement, a constant moisture supply, because the worst thing for a rhodo is to have water collecting around the roots. The number-one reason rhododendrons die is root rot—caused by sitting in water. You need a location with some moisture-retaining abilities coupled with good drainage. I always add plenty of fine fir or hemlock bark mulch and sand to each planting hole, and this has resulted in great success. Another tip: always plant rhodos on the shallow side. Don't plunge those roots deep into the soil.

Everyone assumes rhododendrons love shade, when in fact most varieties stay far more compact, look more attractive and bud up better in a sunny location. The rule of thumb is: the smaller the leaf, the more sun they tolerate; the larger the leaf, the more shade they can take. Filtered sunlight is ideal, because it prevents those broad leaves from drying out, especially during hot weather. Heavy shade is only suitable for large-leafed types and will cause most others to stretch for the light and become gangly.

Rhododendrons vary in hardiness, but keep in mind that fierce winter winds or an exceptionally hot location will cause dessication and subsequent burning of foliage. We all have to experiment a little to find the right plant for the right location and the best performance.

General care for rhododendrons is quite minimal if they have a good growing location, but a light mulching of about 8 cm (3 inches) of fir or hemlock bark mulch will help cool those shallow roots in summer and protect them from rapid freezing in winter.

Proper fertilization will help plants attain good foliage colour, flower better and withstand cold winters with less damage. Remember, it takes a lot of energy to put out those masses of flowers we all love to see. A

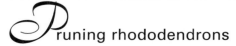runing rhododendrons

If plants are too leggy, they can be pruned back early in spring when the new growth begins. It is best to keep young plants in check by pinching non-flowering shoots back to the next node. Older plants can also be pruned into old, hard wood, but remember to cut on an angle just above a node or swelling on a stem. The harder you prune, the longer it takes the plant to bounce back. One winter a severe cold snap froze several less hardy varieties at our gardens to the ground, and it took three years for them to grow back into nice, 1.2- to 1.5-m (4- to 5-foot) budded plants.

When pruning, try to maintain the shape of the plant by clipping in an oval shape, but don't overdo it. Too many small, weak branches make for a weak plant that will break under heavy snowfall.

It's also a good idea to carefully clip out spent flower trusses. New buds are just below the trusses, and giving them more room encourages better growth and development. Remove the blossom trusses with a quick snap of the hand or with clean, sharp pruning shears. Rhodos do look unsightly after flowering, so a good cleanup also tidies your garden.

quality rhododendron fertilizer, with a 10-6-4 ratio, is ideal, and if the foliage is pale or yellowing, a shot of micronutrients, including iron, will quickly bring back the colour. A product that acidifies the soil and provides micronutrients will rejuvenate pale, tired-looking leaves, but it should be used in conjunction with a rhododendron fertilizer. If you're having difficulty getting blossoms to set, a small amount of sulphate of potash (0-0-50) applied in mid-June really helps.

If leaves are chewed and eaten along the edges, the culprit is probably root weevils. These insects live near the roots of the plant during the day and crawl up the stem to feed on the leaves at night. The best organic method of controlling these pests is to apply a sticky substance called Tree Tanglefoot to the plant's stems about 30 cm (12 inches) above the ground. Wrap masking tape in an 8-cm (3-inch) band around the stem and put the Tanglefoot on the tape. Weevils are caught as they attempt to climb the stem.

Rhodos are versatile plants that can fill many roles in the garden: specimen, background or screening plants, low borders, groundcovers or underplantings beneath trees. But it's important to consider more than just the size of the plant. Think about the time of blooming and the appearance of both new and mature foliage. Also consider the shape and growing habit of the plant. They only bloom a short while, so these other factors are important during the rest of the year.

Dwarf rhododendrons

Early = March to early April   Midseason = mid- to late April

Name	Season	Zone	Flower	Comments
'April Gem'	early to midseason	5	double white	A compact, mounding plant with a slight perfume and pleasant olive foliage.
'April Rose'	early	5	double rose-pink	Dark mahogany-red foliage contrasts well with the flowers.
'Blue Diamond'	early to midseason	6	small, intense blue, all along the stem	Small leaves, great shape. Often used as a border.
'Bob's Blue'	midseason	6	abundant, electric blue	A beautiful semicompact plant with small blue leaves. Ideal for small gardens.
'Dora Amateis'	early to midseason	6	white	A low-growing, tidy plant with deep green foliage. Best in full sun. Spicy fragrance. Great for mass planting.
'Fantastica'	midseason	6	light red with creamy pink centre	Superb dark green foliage and a nice, round habit.
'Ginny Gee'	early to midseason	6	small, pink dappled with white	Very small leaves; almost a groundcover.

Dwarf rhododendrons, cont.

Name	Season	Zone	Flower	Comments
'Hardijzer's Beauty'	early	6	magenta-pink	Ideal for small beds and borders.
'Lori Eichelser'	early to midseason	6	cherry-pink, bell-shaped	Very tight growth habit with deep jade-green foliage.
'Molly Ann'	early and midseason	6	rose	Upright plant with roundish leaves. Looks good year-round.
'Patty Bee'	early to midseason	6	clear soft yellow	One of the best dwarf yellows.
'Purple Gem'	early to midseason	5	small, deep purple	Blue-foliaged compact plants have fragrant leaves.
'Ramapo'	early to midseason	5	pink-violet	Very compact dwarf that does well in sun or part shade. Foliage changes with seasons.
'Rose Elf'	early	6	abundant, orchid-pink	Very compact, mounding plant.
'Scarlet Wonder'	midseason	6	brilliant red	Low mounding habit, looks good all year.
'Shamrock'	early	6	chartreuse	Gorgeous if combined with the purple blues of 'Ramapo' or 'Moerheim Beauty'.

Medium-sized rhododendrons

Early = March to early April   Midseason = mid- to late April   Late = May and later

Name	Season	Zone	Flower	Comments
'Anna Kruschke'	late	5	lavender-blue to reddish purple	Nice-looking, compact plant with lush, dark green foliage. Sun tolerant.
'April Glow'	early	5	rosy pink	Beautiful copper-red new growth.
'Grace Seabrook'	early to midseason	6	blood-red	Outstanding foliage. A great-looking plant.
'Halfdan Lem'	midseason	5	huge, bright red trusses	The contrast of the red flowers with thick, deep green foliage make it outstanding.

Name	Season	Zone	Flower	Comments
'Honourable Jean Marie Montague'	midseason	6	brilliant red	Vibrant buds and thick, attractive foliage make it look outstanding all year round. My favourite red.
'Ilam Violet'	midseason	6	deep violet	Bronze winter foliage. Ideal to plant with deciduous yellow azaleas. Magnificent.
'Ken Janeck'	midseason to late	4	dark pink edges fading to soft pink	Leaves have attractive fuzzy brown undersides.
'Lee's Dark Purple'	midseason to late	5	dark purple	Wavy, dark, attractive foliage.
'Lem's Cameo'	midseason	6	apricot-cream and pink	New growth is bronze-red, maturing to shiny green. Outstanding.
'Mist Maiden'	midseason	4	misty pink	Large leaves have beautiful fawn coloured undersides. Silvery new growth.
'Mrs. Furnival'	midseason to late	6	dark purple with black blotches	Deep green foliage. Queen of the purples.
'Nancy Evans'	midseason	6	bronzy yellow	Attractive compact, rounded habit. A winner.
'Rocket'	midseason	6	vibrant pink blotched scarlet, frilled	Midsized, somewhat ruffled in appearance. A good-looking plant.
'Song'	late	6	orange	Attractive olive foliage grows three times longer than it is wide.
'Trude Webster'	midseason	6	huge, clear pink trusses	Foliage is large, wide and slightly twisted and matches the look of the trusses. Magnificent.
'Unique'	early to midseason	6	pink, then buttercream	Smooth, rounded leaves on compact attractive plant.
'Yellow Petticoats'	midseason	6	clear deep yellow, frilly	An attractive, compact plant with good, dark green contrasting foliage.

Tall rhododendrons

Early = March to early April   Midseason = mid- to late April   Late = May and later

Name	Season	Zone	Flower	Comments
'Anna Rose Whitney'	midseason to late	6	rosy pink, large trusses	Huge and fast-growing with large, 20-cm (8-inch) leaves.
'Gomer Waterer'	midseason to late	5	pink buds open to pure white	Large, deep green leaves make it an outstanding foliage plant. One of my favourite whites.
'Pink Walloper'	midseason	6	huge, satiny pink trusses	Very large leaves with reddish leaf stems give the look of a large, luxurious plant.
'Red Walloper'	midseason	6	rose-red, then fade to pink	Nice, strong foliage; a great-looking large plant.
'Taurus'	early to midseason	7	deep red with slight black speckles on petals	Fabulous shrub with deep green, sturdy leaves and buds that have a reddish hue in winter.

Conifers

Evergreen conifers have a new role to play in today's gardens. Virtually gone are the massive junipers, large cypress and cedar trees that were once so popular. We still need the colour, texture and form of conifers, but we want them to blend with deciduous shrubs for a show in winter and complement broadleaf evergreens, such as heather, azaleas and compact rhododendrons.

Compact conifers provide contrast and winter interest, and weeping and odd-shaped conifers are wonderful focal points in containers. Conifers also make the best hedges for screening and privacy.

Best bets—landscape feature trees

Even in small garden spaces, we need a bit of flair, and there are many conifers with interesting and beautiful shapes that can inspire a landscape. Weeping forms can soften and create interest in an otherwise ho-hum garden. Some conifers need more space than others, but there is definitely a place for these plants in the landscape.

Hollywood juniper (*Juniperus chinensis* 'Kaizuka'). Zone 5. Grows to 3 m (10 feet). I love the windswept appearance of this semi-twisted, blue-green, upright juniper. It looks fabulous in containers and makes an interesting focal point. Wonderful for seaside planting.

Korean fir (*Abies koreana*). Zone 4. Grows to 10 m (33 feet). A slow-growing, very symmetrical tree with fragrant needles, silver on the underside. Attractive purple cones are borne even on young trees. A great landscape specimen.

Sargent's weeping hemlock (*Tsuga canadensis* 'Pendula'). Zone 4. Grows to 2 m (7 feet). A moderately fast-growing form, with long, weeping branches.

Slender hinoki cypress (*Chamaecyparis obtusa* 'Gracilis'). Zone 5. Grows to 4 m (13 feet). A wonderful slender form with very rich green foliage that has a twisted look. It's a great focal point plant, works well in containers and softens the corners on any building.

Weeping blue Atlas cedar (*Cedrus atlantica* 'Glauca Pendula'). Zone 6. Grows to 3.6 m (12 feet). The main stems and branches hang down in a weeping form, and the needles are a steely blue. I've seen this plant used as a novelty fence, staked as a weeping tree and trained beautifully over embankments. It's a good container plant as well.

Weeping nootka cypress (*Chamaecyparis nootkatensis* 'Pendula'). Zone 6. Grows to 8 m (26 feet). A tall, slender, elegant tree with some branches that gently push out like arms and some that hang straight down. Wind in the branches creates a lovely flowing effect.

Weeping sequoia (*Sequoiadendron giganteum* 'Pendulum'). Zone 6. Grows to 8 m (26 feet). This wonderful specimen has a twisted trunk with long branches hanging straight down. It looks like a giant Grinch.

Weeping white spruce (*Picea glauca* 'Pendula'). Zone 3. Grows to 5 m (17 feet). One of the most graceful weeping spruces with blue green foliage and a graceful, open habit. Long, pitchy pine cones in fall.

Best bets—compact conifers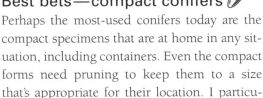

Perhaps the most-used conifers today are the compact specimens that are at home in any situation, including containers. Even the compact forms need pruning to keep them to a size that's appropriate for their location. I particularly like varieties with an unusual colour, an interesting form or striking winter colouring.

Alberta spruce (*Picea glauca albertiana* 'Jean's Dilly'). Zone 4. Grows to 1 m (3 feet). All Alberta spruce are attractive when young, but as larger plants they look a little sloppy. 'Jean's Dilly' has a compact, tight form that looks great even as an old-timer.

Blue star juniper (*Juniperus squamata* 'Blue Star'). Zone 5. Up to 50 cm (20 inches) tall. Very striking, blue compact form that

Pruning conifers

Although you can prune conifers almost any time of year, I always like to prune them just before the new growth begins in early spring. This gives them a more compact form. Regular pruning, especially when the plants are young, will keep them looking great in your landscape far longer. Faster-growing conifers can be pruned again in the first week of July, when they go into a brief summer dormancy before they take off again in late summer.

Three notable exceptions are pine, spruce and fir. If you prune them in winter, or before the buds develop, you cut off a whole year's growth. Prune them just when the "candles" (pine) and buds (spruce and fir) have opened and the new growth is well under way. This will allow new growth to form along the pruned branches for a much fuller and more compact growth habit next year. Never prune in midsummer hot spells, when removing protective growth will cause inside branches to burn; wait for rainy, cloudy weather.

resembles a star when mature. Needs exceptional drainage.

Creeping juniper (*Juniperus horizontalis* **'Mother Lode').** Zone 4. Grows to 30 cm (12 inches). One of the most brilliant gold carpet junipers.

Dwarf globe blue spruce (*Picea pungens* **'Glauca Globosa').** Zone 3. Grows to 1 m (3 feet). The intense blue of this plant is unlike any blue in the garden. It stands out beautifully any time of the year.

Dwarf hinoki cypress (*Chamaecyparis obtusa* **'Nana Gracilis').** Zone 6. Grows to 1.2 m (4 feet). The swirled and twisted branches of this rich green Japanese cypress make it a great compact focal point. Needs superb drainage to prevent root rot.

Dwarf mugo pine (*Pinus mugo* **var.** *pumilo*). Zone 2. Grows to 1.5 m (5 feet). This low, spreading pine is a natural in rockery and alpine gardens.

Dwarf Serbian spruce (*Picea omorika* **'Nana').** Zone 5. Grows to 1 to 1.5 m (3 to 5 feet). A beautiful compact spruce with blue-green needles, silver on the underside. Has a very dense, broadly conical form.

Gold thread cypress (*Chamaecyparis pisifera* **'Mops').** Zone 4. Grows to 1 m (3 feet). This compact form with intense gold colouring adds life to any garden, especially in winter. It softens stiff plantings with its weeping habit and droops nicely over the edge of containers.

Low spreading hemlock (*Tsuga canadensis* **'Bennett').** Zone 5. Grows to 1.2 m (4 feet). Very slow-growing, dwarf shrub with a spreading habit and dense growth. The chartreuse buds of the new growth look particularly nice.

Rheingold cedar (*Thuja occidentalis* **'Rheingold').** Zone 3. Grows to 1.2 m (4 feet). A low, round cedar with unusual orange foliage that stands out brilliantly in winter. It's great planted with winter heather and dark-coloured conifers.

Spreading golden English yew (*Taxus baccata* **'Repandens Aurea').** Zone 5. Up to 50 cm (20 inches) tall with a 1.2-m (4-foot) spread. The prostrate habit of this spreading yew is perfect over stones as a groundcover.

Best bets—hedges

We all appreciate a little privacy, and certain conifers make wonderful year-round screens and hedges. Hedges allow privacy and make a good barrier, keeping out neighbours' pets and defining the boundaries of the property. In addition, they can deflect cold winter winds and cool summer breezes that make sitting out on the patio in the evening a little chilly.

One of the prime considerations when selecting hedging plants is maintenance. Consider this in terms of both tomorrow and 10 years from now. Fast-growing evergreens need one or two trimmings a year and can ultimately grow to somewhere in the neighbourhood of 30 to 40 m (96 to 128 feet).

Slower-growing, more compact trees may

Tips on planting hedges

Prepare a planting bed by taking off the sod and digging down at least 36 to 60 cm (14 to 24 inches) and 60 cm (24 inches) wide. Make sure this long bed is well prepared with good soil. To loosen heavy soils, add fine fir or hemlock bark mulch and sand. Stones and gravel should be removed and replaced with good soil. Do not add manures as they are alkaline, and all these trees are acid loving. Large trees should be loosely staked until they root. This preparation is critical to the establishment and ultimate growth of the plants.

be the best long-term choice. When you plant a hedge, always consider your neighbours. Plant the trees far enough into your property to prevent them from encroaching on the yard next door. Take a look at mature trees in your neighbourhood—this will give you a sense of how wide they can become.

Heavy winter snows can damage hedging plants, so be sure to select varieties that shed snow or can be easily pruned to do so.

Brown's yew (*Taxus media* 'Brownii'). Zone 5. Grows to 1.2 to 2.4 m (4 to 8 feet). A broad columnar tree with upright branches. Requires little pruning and does well in lightly shaded areas. This male variety does not produce seed pods, which are toxic. Plant 40 to 60 cm (16 to 24 inches) apart. Grows 30 to 40 cm (12 to 16 inches) per year.

Emerald cedar (*Thuja occidentalis* 'Smaragd'). Zone 3. Grows to 3 m (10 feet). One of the richest green cedars with a narrow form and attractive swirled branches. Has good winter colour and does not show unattractive winter seed pods. Needs little pruning. Plant 60 to 70 cm (24 to 28 inches) apart, centre to centre. Grows 30 to 50 cm (12 to 20 inches) per year.

Excelsa cedar (*Thuja plicata* 'Excelsa'). Zone 5. Grows to 6 to 60 m (20 to 200 feet). A selection of *T. plicata* (western red cedar) that is much more compact and has better winter colour. It needs to be pruned at least once a year to keep it narrow. Ideal for a tall windbreak, but will burn in fierce winter winds. Plant 1 to 3 m (3 to 10 feet) apart, centre to centre. Grows 50 to 90 cm (20 to 36 inches) per year.

Hardy pyramid cedar (*Thuja occidentalis* 'Brandon'). Zone 2. Grows to 3.6 to 9 m (12 to 30 feet). A very hardy selection of the pyramid cedar, *T. occidentalis* 'Fastigiata', with similar growing habits.

Irish yew (*Taxus baccata* 'Fastigiata'). Zone 6. Grows to 3 to 9 m (10 to 30 feet). A deep rich green hedging plant with a very narrow form. We've used them as a hedge around our rose garden for 20 years with wonderful results. Plant 40 to 50 cm (16 to 20 inches) apart, centre to centre. Grows 20 to 30 cm (8 to 12 inches) per year.

Leyland cypress (*Cupressocyparis* × *leylandii*). Zone 6. Grows to 15 to 30 m (50 to 100 feet). Growing 1.5 m (5 feet) in one year, it is often known as the world's fastest-growing conifer. Its blue-green foliage is attractive, and it will tolerate shade and ocean conditions. Needs constant pruning when young to form a compact, handsome hedge. Will burn in cold winter winds. Plant 1 to 3 m (3 to 10 feet) apart, centre to centre.

Pyramid cedar (*Thuja occidentalis* 'Fastigiata'). Zone 3. Grows to 3.6 to 9 m (12 to 30 feet). Narrow pyramidal cedar with good summer colour, turning dull green in winter. Does tend to show unsightly seed pods when under stress. Plant 60 to 70 cm (24 to 28 inches) apart, centre to centre. Grows 40 to 50 cm (16 to 20 inches) per year.

Fruit and Nut Trees and Small Fruits

Nothing beats the flavour of fruit fresh from your own garden. With new, disease-free varieties and space-saving compact forms, fruit and nut trees can be a valuable addition to any garden, no matter how small. One of the main advantages is that you can have varieties that might not be commercially viable because they don't meet storage and handling requirements. (It amazes me that it's more important for a piece of fruit to look good on the shelf than for it to taste good.) Another plus is that you can grow pesticide-free fruits and nuts, a big factor for people concerned about what they're serving their families.

Fruit and nut trees also play an important role in the landscape. There are new varieties that make excellent privacy barriers, shade trees or "living fences." For a bonus, most fruit trees are quite spectacular when they flower in spring. All these other benefits fall by the wayside, though, when you pick that first, perfectly ripe fruit of the season and take a bite.

Some people don't plant fruit and nut trees because they think they're a lot of work. I can't say they're maintenance free, but it's not really too big a job to take care of a few well-chosen trees. These are the major tasks.

- Dormant spraying to keep the bark free of insects and diseases.
- Pruning to keep the tree free of dead and diseased wood, maximize fruit production and maintain a manageable size.
- Summer pruning to control water sprouts.
- Some fungicide spraying to prevent diseases like scab, mildew and European canker.
- A little leaf raking in the fall.

The good news is that so many new varieties and growing techniques have been developed over the past few years that many of these chores have been minimized. In the following pages you will find information on the best varieties as well as techniques to keep care to a minimum.

Fruit Trees

Grafted trees

Fruit trees are most often a blend of separate parts that are bud-grafted, or joined together, to create a better plant. A typical fruit tree is grafted to a rootstock that determines factors such as the tree's final size, the type of soil in which it will grow best, its strength and the age at which it will bear fruit. But there are considerable differences between the various types of fruit trees. For example, "dwarf" is a relative term and can mean a fruit tree anywhere from 3 to 6 m (10 to 20 feet) tall—a huge variation in size! As a rule, the standard trees—either grafted or from seedling—can take from 7 to 10 years to produce fruit, and usually grow from 8 to 12 m (25 to 40 feet). Dwarf trees usually produce fruit in half the time and will be 25 to 50 percent as big as their standard counterparts.

- Most apple rootstocks were developed in England at the East Malling (M Series) and Merton (MM Series) research stations. The lowest-growing apple trees are generally on an 'M9' rootstock, producing a 2.4- to 3-m (8- to 10-foot) tree-bearing in three to four years. Although ideal for espaliers, trees on 'M9' rootstocks require staking or other support because the roots are not strong enough to support the crown. The most common of the home garden dwarf tree stocks is 'M26'. It's strong and will result in a final tree size of about 3 to 3.6 m (10 to 12 feet). A semidwarf rootstock is 'M7'; it has good vigour and grows to about 3.6 to 5.4 m (12 to 18 feet) in height and spread.

*P*omes and stones

Most fruit trees can be grouped into two categories: pome fruits and stone fruits. Fruits in each category have cultural similarities. The two pome types—apples and pears—have similar pest problems and growing characteristics. The stone fruits—plums, cherries, peaches, nectarines and apricots—also have similar needs and problems.

All of these rootstocks do best in fairly fertile soils, so if you have poor, sandy soil, look for either 'M4' or 'M27', which are more suitable for those conditions.

- Nearly all pears are grown on 'Quince A' rootstock, producing trees that grow to about 3 to 6 m (10 to 20 feet).
- Cherry trees grow the largest of all the stone fruits, reaching up to 18 m (60 feet) or more. 'Mazzard' rootstock will reduce that to 80 percent of a seedling tree, while 'Colt' rootstock keeps growth to about 4.5 to 6 m (15 to 20 feet). The 'Gisela' rootstock from Germany provides the first opportunity to create a true dwarf cherry, about 40 percent of the size of a standard tree. It also allows the tree to produce within a year or two.
- Most plums, peaches, nectarines and apricots are grown on the dwarfing rootstock 'St. Julien A'. Trees will grow 2.4 to 4.5 m (8 to 15 feet), depending on variety.

It may seem complicated; but once you know even a little bit about rootstocks, it will help you select the right tree for your landscape. Every year there are more and more varieties available, so it's hard to recommend any one variety as the be-all, end-all fruit tree of its type. If you're dealing with a newer variety, ask the following questions before you buy.

- How big will this tree get?
- How much horizontal space does it need?
- How much maintenance does this tree require?
- How does it fit in with the growing conditions in my garden, such as sunlight, climate and soil conditions?

Decide what type of fruit you like best, measure the area where you want to plant and talk to the retailer to make sure you won't get an unpleasant surprise down the road.

Pollination

Very few fruit trees are self-fertile. This means you need two varieties to get fruit on either. If you only plant one variety, and there is not another for pollination in your area, you will not get fruit. I've seen this set back many fruit-eating plans by a couple of years, so when you choose a tree, make sure you find out if it is self-fertile. Sometimes you can get one tree that has been grafted with two or more varieties.

To make things even more challenging, you have to get varieties that are able to pollinate each other—don't assume any combination will work. They must bloom at the same time. Triploids (apples with an odd set of chromosomes) don't produce enough pollen to pollinate another tree. Examples include 'Belle de Boskoop', 'Bramley's Seedling', 'Gravenstein', 'Jonagold', 'King' or 'Mutsu'.

Location and planting of fruit trees

Trees can live several generations, so spend some time choosing the right location. At the top of the list is deep, well-drained soil. Wet soils are perilous to all fruit trees. Full sun is also essential to ripen fruit, ensure good production and minimize disease problems. Watch out for "frost pockets" in low-lying areas. A late frost can destroy an early flower crop and subsequent fruiting.

Planting is best done in late winter when a fresh shipment of fruit trees arrives at your favourite garden centre. Even though the tree is dormant, the roots will start to establish over winter. This will result in much better growth the first season. Spend the time to follow these tips on preparing the planting site—the effort is well worth the rewards down the road.

- Dig a generous hole one and a half times deeper and wider than the root ball.
- Fill the bottom of the hole with well-rotted compost and some bone meal to help the roots along. Fork it in to thoroughly blend it with the existing soil. Bark mulch is a terrific addition if the soil is heavy or poorly drained.
- Drive a strong stake into the ground if the tree will need support. You only need to stake trees on very dwarfing rootstocks. Staking unnecessarily can result in a weaker tree, so ask if your tree needs staking when you're buying it.
- In fall and early spring, dormant trees may be available "bare root"—without soil on their roots. This gives you an opportunity to see the tree's root system. Choose trees with well-developed roots that branch out into smaller feeder roots. Prune back any broken roots, and cut back extra-long root stems for a balanced flare of roots. Dip the roots in a mud slurry containing liquid root booster for a faster start (see chapter 1 for more information).
- Place the tree in the hole and keep the bud union 5 to 10 cm (2 to 4 inches) above ground level on grafted trees.
- Many garden centres get a fresh shipment of fruit trees each spring that are potted into pressed-paper fibre pots. Always lift these trees by the pot—they're not completely rooted until summer.

Set the pot into your prepared planting hole. When the hole is almost filled with soil, peel back the top several inches of the fibre pot. Don't leave the fibre pot sticking up above the soil level—it will act as a wick, drawing water away from the roots.

Choosing a healthy tree

When buying, look for a relatively straight stem, with three or more branches radiating outward in different directions. One-year-old trees may be sold as nonbranched "whips." Avoid trees with branches that attach to the trunk at a narrow angle—a wide branch crotch is stronger. Look for clean trees without moss, algae or splitting bark.

- Sometimes fruit trees are sold "balled and burlapped" (b&b). Rather than being bare-rooted, they are dug from the field with a ball of soil that is wrapped in burlap to secure it.

  When planting b&b trees, set the root ball in the planting hole. When the hole is about half full, carefully cut the string around the neck of the tree and fold the burlap away from the top of the ball. As you continue filling the hole, cover the burlap, which can also act as a wick.

  Be cautious with b&b trees, especially if the root ball seems soft. Sometimes, particularly with poorly treated root balls, it is better to plant the entire ball and cut the string only after the root ball has been firmed in with your planting soil.
- Tamp down the soil and thoroughly water the roots to eliminate air pockets.
- Use a watering can to add liquid root starter to activate the root system.
- Add an 8-cm (3-inch) layer of mulch on top to hold in the moisture and protect the new roots from frost or heat.
- In spring, loosen up the soil carefully with a digging fork to allow more oxygen into the soil.

- Tie the tree to a stake if you're supporting it. Use something that will not bite into the bark, such as a piece of old garden hose. Allow a bit of slack—gentle movement will make the trunk stronger.

Care and feeding

Water deeply during long dry spells, particularly when the fruit is sizing up. Use a soaker hose around the drip line of the tree at least once a week. This helps prevent fruit drop and your fruit will taste sweeter.

Work some well-rotted manure or compost into the top layer of the soil each spring. Add some 10-14-21 or a similar formula fertilizer, including micronutrients, to help the roots develop and encourage the tree to produce more fruit. Apply it in early spring when the tree starts to show the first signs of life, and again in midsummer when the fruit is sizing up.

Pruning and training

If you look at a successful orchard, you will see that the trees have not only been pruned to shape, but also trained. Always remember, pruning stimulates *more* growth.

Some pruning is necessary, even with train-

Planting a bare root tree

The old soil mark on the stem should be level with the bottom of the board. The board is used to ensure correct planting depth.

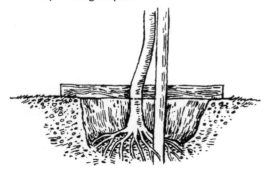

Gently add planting mixture around roots. Use your hands to press down mixture.

ing. Prune after the leaves are off, any time from late November until just before the sap begins to flow in late winter. Early spring is ideal because there are fewer disease organisms in the air. Never prune if it's raining or very frosty. Always prune outside the branch collar (the thickened area where the branch joins the main stem). You don't need to use pruning paint on areas where you've cut. It's been shown to impede the healing process.

Apples, pears and plums are often pruned to an "open vase" shape, which allows fruit to develop on the inside of the trees, on the tips and on the outward-growing branches. For this shape, cut the leader (main growing stem) at the desired height. Prune off all but three to five healthy, well-placed branches radiating out from the main stem. The branches should be high enough off the ground so you can walk or work underneath the tree without whacking your head on them.

Once you have pruned off the main branches, cut out any branches left in the centre and any inward-growing branches. Next,

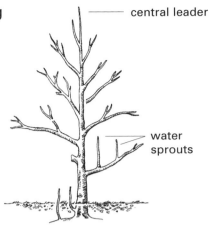

Pruning a tree

cut out all the upward-growing branches, leaving the tree looking like an open vase, with the main branches radiating out. The remaining branches should be pruned back each year by about one-third, keeping this vase formation intact.

A heavy pruning each year will result in masses of water sprouts shooting out in all directions. To minimize this problem, once you have developed the open vase shape described above, switch to a training program

Pruning sweet cherries

Sweet cherries present a slightly different problem. They are perhaps the most vigorous of all fruit trees and resist any attempt to be pruned in an open vase form. Sweet cherries tend to have a very strong stem or leader, so they require "central leader pruning." This means cutting back the main stem each year to control the rate of growth and, at the same time, cutting back the outward-growing branches, creating a pyramid-shaped tree. If there are two or three strong central stems, it may be a good idea to remove them to avoid competition and to thin out the tree.

The central leader will develop two or three branches after pruning, and the one that grows into the strongest and most upright stem should be treated as your central leader next season. The other stems can be removed or left, depending upon their growth habit. If they can be trained in an outward direction, cut them back next year along with the other branches. Try not to let the centre of the tree become cluttered.

rather than a heavy pruning regime and only prune a portion of the tree each year. This will minimize the need for massive pruning and greatly reduce the number of water sprouts you have to deal with. Horizontal branches will be the best fruit-bearing stems on the tree.

The easiest way to train trees in a horizontal form rather than an upright pattern is to tie weights to the ends of the main branches. The goal is to create a tree with as much horizontal growth as possible to encourage the development of open spaces and fruit-setting branches that will stay low to the ground, in easy reach for harvest. This training is best done on young trees, but older trees can be modified as well. You want to give them a weeping willow look until the branches hold the horizontal pattern on their own. I often fill small sandwich bags with sand or soil and carefully tie them to the ends of branches to achieve this.

Another benefit of training as opposed to pruning is a reduction of sucker growth. The more you prune, the more the tree will put out suckers and water sprouts in the spring. Remove this growth all year round as it appears.

Espalier

Espalier is an old technique that is becoming popular again. Apples, pears, plums, peaches, nectarines and apricots are all suited to espalier, which means training the tree to grow flat against a structure or frame, usually in a decorative pattern. It can be used in small spaces, such as against a wall, or to create an attractive and productive fence. Start with either one-year-old whips or very pliable two-year-old trees on the most dwarfing rootstock you can find. The idea is to cut the trees back to less than a metre (3 feet) on a single stem to create what is called "feathering." This is the development of low side branches that can be trained along a wire or wall to grow horizontally or in a variety of shapes. These branches

Espalier

In summer, train growth along the canes.

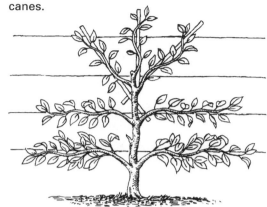

can then produce an amazing amount of fruit in a very small area.

The "cordon" (ropelike) style is the most popular, but there are others, such as "palmetta" (palm-shaped). A visit to a botanical garden with a number of different fruit tree forms is an ideal way to get some inspiration. When trained fruit trees are in blossom they match the most exotic flowering trees for beauty.

Both pruning and espalier are complicated tasks, and I suggest you take a look at a book devoted exclusively to these subjects. Pruning and training can be a real art form and a little bit of effort can create showcase trees. So remember when you're out there with your saw and shears—it's not work, it's art.

A Fruit Tree Sampler

After 30 years of helping people select fruit trees for the garden, I've learned a few things. If they have their heart set on a certain apple variety, for instance, because of the flavour, tartness, keeping qualities, cooking attributes or just for good old sentimental value, I always try to match them up with that tree. If their choice has some particular problems, I always point that out so they can take appropriate measures, or make another choice.

Many folks make their pick based on tasting fruit from the store and deciding it would be great to have that variety in the back yard. What they often don't realize is that the apple or peach or plum is from a different climate and probably won't do well in their neck of the woods. That's when I try to steer them to a similar tree adapted to local growing conditions. As a rule I try to recommend varieties that are low maintenance—easy to grow and reliable fruit producers with the least care. I realize most people are not experts and just want to enjoy the pleasure of having their own fresh fruit.

enovating out-of-control trees

Sometimes we move into a house and inherit fruit trees that have been untouched for many years. At this point we have to decide whether it would be better to replace the tree with a new variety or try to revive the old one. Older trees tend to have lower production after 20 or 30 years. Sometimes, however, the tree is valued for what it adds to the landscape as much as for the fruit it produces. Here's what you have to consider when it comes to rehabilitating an older, neglected tree.

1. Scrape the bark of dead-looking branches with your fingernail or a knife. If there is green underneath, it tells you the tree is still alive.
2. Check out the growth buds; pick one off to see if it's still plump and green.
3. Get rid of dead wood and prune back to where there is life in the tree.
4. Rebuild the soil. Carefully dig around the roots to loosen the soil. If it is heavy and drainage is poor, add fir or hemlock bark mulch. Next add 10 to 15 cm (4 to 6 inches) of well-rotted manure.
5. Fertilize around the drip line of the tree with something like 10-14-21. Make sure the fertilizer includes micronutrients in case the soil is deficient in some way.
6. It also often helps to add dolomite lime around the base of older trees. This makes the soil less acidic and adds magnesium.

Older trees are more susceptible to algae, fungus, lichen and insects that burrow into the crevices. I recommend spraying the tree with an organic spray such as lime sulphur and dormant oil. Apply it three times over the winter: November, December and January. Ask at your local garden centre for details about using the spray.

Pruning a neglected tree
Large cuts should be made close to the remaining limbs.

Apples (*Malus domestica* varieties)

Apples can be divided into two main types: eating and cooking apples. There are hundreds of varieties to choose from and each year dozens of new ones are introduced. But old ones still remain popular and many are making a comeback. 'Cox's Orange Pippin', for example, was first introduced in England in 1825, but is still popular today.

When choosing a tree, think about four key criteria.
- taste
- disease tolerance
- tree size relative to space available in your garden
- keeping and storing qualities of the fruit

Most home gardeners select dwarf trees on the self-supporting 'M26' rootstock. Some folks espalier their trees along a fence or wall for wonderful results. Colonnade types are also gaining in popularity. These trees grow in a narrow column and can be very useful in a restricted space.

Apple varieties are often classified according to their ripening time: early, midseason or late. Early varieties ripen by mid-September and their starch converts to sugar more rapidly, meaning they soften quickly and generally lose their quality in a shorter time. Late varieties ripening from mid-October onward remain hard, and if eaten too soon, taste a little starchy. As they continue to metabolize they become sweeter. Midseason varieties split the difference, becoming sweeter earlier but not lasting as long as the late varieties.

- 'Akane' (Prime Red). Attractive red eating apple. Moderate scab resistance. Midseason ripening.
- 'Belle de Boskoop'. Dull, rough skin, yellow-bronze to red. Juicy and tart. Midseason ripening.
- 'Braeburn'. Green-shaded dark red. Stores well. Good flavour.
- 'Bramley's Seedling'. Large-sized fruit, green flushed with red. Late ripening. Classic cooking apple.
- 'Cox's Orange Pippin'. Medium-sized, round, yellow-orange fruit with excellent flavour. Ripens mid-September.
- 'Discovery'. Large, yellow-red apple with excellent flavour. One of the most popular varieties in England.
- 'Egremont Russet'. Rough golden skin on a great dessert apple. Makes a good espalier.
- 'Elstar'. Similar to 'Golden Delicious' and 'Cox's'. Unique flavour. Stays white when cut, excellent for salads.
- 'Florina'. Light, bright red. Outstanding flavour, ranks with 'Jonagold' and 'Elstar'. Very scab free and disease tolerant. Ripens mid-October.
- 'Freedom'. Similar in colour to 'Jonagold', with red cheeks. Late fall ripening. Good keeper, scab resistant. Large, crisp, juicy fruit.
- 'Fuji'. Excellent green winter apple with red stripes. Outstanding flavour, stores well and is the most popular Japanese variety. Ripens late October.
- 'Fuji Red Sport'. Yellow-green with pink stripes. Ripens mid- to late October. Stores

---

Picking and storing apples

You can tell when apples are ripe and ready for picking by colour and size. You can test by lifting and tilting them to 90 degrees. If the stem comes off the spur easily, they're ripe.

Store apples in a cool, dark area just above freezing. Keep them in plastic bags. The citric acid they give off helps to preserve them.

well. 'Gala' or 'Granny Smith' can be used as a pollinator.
- 'Gloster'. Similar to 'Red Delicious'. Huge, red apple known in Germany as the "Christmas apple." Good keeper, quite sweet when fully ripe.
- 'Golden Reinette'. Also known as 'Triumph de Boskoop'. Medium-sized golden yellow fruit with slight russeting. Juicy and flavourful.
- 'Golden Russet'. Medium-sized, brown-yellow fruit. Sweet and flavourful, good for sauce. Late ripening.
- 'Granny Smith'. Large, round green apple. Not suited to coastal areas. Fair quality. Good for pies. Late ripening.
- 'Grimes Golden'. Medium-sized, round, yellow fruit. Firm flesh. Sometimes mealy, but good flavour. Midseason ripening.
- 'Jerseymac'. Summer 'McIntosh'. Ripens early. Good flavour.
- 'Jonafree'. Crisp, firm, juicy. Good dessert quality. Excellent disease resistance. Mid-season ripening.
- 'Jonagold'. Medium-sized yellow fruit with red stripes. Crisp, good flavour. Late ripening.
- 'Jonathan'. Large, bright red fruit with a tart, sweet flavour. Good pollinator. Late ripening.
- 'King'. Very large, midseason to late apples. Red and yellow striped. Good for eating fresh and makes great cider.
- 'Liberty'. One of the finest disease-free varieties. Red fruit similar to 'Spartan'. Juicy. Ripens mid-September. Best scab- and mildew-resistant variety.
- 'Lodi'. Green apples, similar to 'Transparent', but larger and more regular bearing. Good for table, sauce or pies. Ripens very early.
- 'McIntosh'. Round, red fruit. Soft flesh. Not recommended for growing in coastal areas. Good for all uses, including apple cider. Ripens mid- to late September.
- 'Melba'. Sometimes called 'Coast McIntosh', with medium-sized green fruit with red stripes. Flesh is white, fine-textured and crisp. Early ripening.
- 'Mutsu'. Crisp, yellow-green fruit. Excellent flavour. Highly rated for coastal areas. Good keeper. Ripens midseason to late.
- 'Northern Spy'. Round, medium-sized red apple. Midseason ripening. Good for cider and applesauce.
- 'Nova Easygro'. New early variety. Ripens three weeks before 'McIntosh'. Medium-large, crisp red fruit. Good disease resistance. Store before eating for best flavour.
- 'Orenco'. Large, red-yellow fruit. Delicious white flesh. Table use only.
- 'Prima'. Newer introduction from the United States. Large, crisp and juicy. Yellow with red stripes. Good fresh quality and a good keeper. High scab resistance.
- 'Priscilla'. Another newer introduction from the U.S. Scab-free, large, dark red fruit. Average flavour. Not a good keeper. Dessert apple.

Best bets—disease resistance

One of the latest trends, especially in wet climates, is the use of disease- and scab-tolerant varieties that produce pretty good apples on trees requiring very little fungicidal spraying. Here are my picks in this category.
- 'Florina'
- 'Freedom'
- 'Jonafree'
- 'Liberty'
- 'Redfree'

- 'Red Delicious'. Large, crisp red fruit. Good flavour, the number-one supermarket apple. Table use. Late ripening.
- 'Redfree'. Outstanding! The best early season coastal variety. Glossy, red fruit. Scab free. Can be stored up to two months.
- 'Red Gravenstein'. Striped yellow-red fruit. Firm, juicy, tart flavour. Favourite coastal apple.
- 'Royal Gala'. Red-striped fruit. Excellent flavour. Ripens late summer. For table only.
- 'Shamrock'. Large, round, green, 'Granny Smith' type. Early variety, ripens late September.
- 'Spartan'. One of the best apples for coastal areas. Small, dark red fruit. Juicy, crisp, delicious flesh. Best for table use. Good fresh or cooked. Midseason ripening.
- 'Sunrise'. Green fruit with red blush. Excellent quality. Ripens mid-August.
- 'Tydeman's Red', or 'Tydeman's Early'. Medium-sized, attractive red fruit. Excellent flavour. One of the best early apples.
- 'Wealthy'. Medium-sized, red-striped fruit. Juicy, crisp and delicious. Stores well.
- 'Winter Banana'. Large, pale yellow with a pink blush. Juicy. Very good flavour. Table use only. Late ripening.
- 'Yellow Transparent'. Earliest coastal apple. Green, tart and flavourful when ripe. Excellent for applesauce and pies.

Best bets

Flavour
- 'Cox's Orange Pippin'
- 'Discovery'
- 'Elstar'
- 'Florina'
- 'Fuji'
- 'Jonagold'
- 'Mutsu'
- 'Northern Spy'
- 'Royal Gala'
- 'Spartan'
- 'Tydeman's Red'

Long-keepers
- 'Braeburn'
- 'Florina'
- 'Fuji'
- 'Golden Russet'
- 'Mutsu'
- 'Northern Spy'
- 'Winter Banana'

Columnar apples

This revolutionary type of apple tree originated in British Columbia's Okanagan region and has undergone 25 years of research, trials and development. The result is space-saving, easy to grow, heavy-cropping garden apples of excellent flavour and quality. They reach a height of 2 to 2.4 m (7 to 8 feet) after five years and top out at a maximum of 4 m (13 feet).

Columnar apples require little or no pruning. If side shoots develop, simply cut them back in winter to leave two buds. They can be planted in flower beds, lawns and tubs, or used as a hedge. They fruit early in their lives and are highly productive. Varieties will cross-pollinate and fruit set is enhanced when more than one variety is planted.

- 'Bolero' ('Emerald Spire'). Shiny green apples with golden blush. Crisp, juicy flavour. Early ripening.
- 'Golden Sentinel'. 'Golden Delicious' type. Medium-sized, flavourful fruit. Scab free. Ripens mid- to late September.
- 'Polka' ('Scarlet Spire'). Bright, red-green apple with a flavour similar to 'Spartan'. Midseason ripening.

- 'Scarlet Sentinel'. 'Idared' or 'Melba' type. Matures mid- to late October. Medium-sized, roundish to oblate shape. Dense, juicy and flavourful. Scab resistant.
- 'Waltz'. Dark, red-green fruit. Sweet and juicy. Good keeper. Late ripening. Purplish pink and white blossoms early to mid-May.
- 'Wijcik'. Superb 'McIntosh' type. Excellent flavour. Good for patio pots.

Apricots (*Prunus armeniaca* varieties)

Self-fertile varieties are best for small gardens. In wet climates, they need to be grown in espalier form against a south- or west-facing wall.
- 'Blenheim'. Round, medium-sized fruit. Gold with red blush. Self-fertile. Good for drying. Good pollinator. Ripens early August.
- 'Harglow'. A newer west coast variety. Medium-sized, orange-coloured fruit. Excellent disease resistance. Ripens mid-August.
- 'Moorpark'. Large, oval, yellow fruit with greenish yellow flesh. Self-fertile. Good for canning and drying. Good pollinator. Ripens early August.
- 'Perfection'. Large fruit with orange yellow skin and flesh. Excellent flavour. Ripens early August. Plant with a good pollinator.
- 'Pui-Sha-Sin'. Large, round, juicy orange fruit with excellent flavour. Ripens late July. Plant with a good pollinator.
- 'Sundrop'. Mid-sized, round, waxy fruit. Good flavour. Self-fertile. Ripens early August.
- 'Tilton'. Small to medium-sized, golden yellow, flavourful fruit with red blush. Self-fertile. Good for canning and drying. Good pollinator. Ripens mid-August.

Cherries (*Prunus* varieties)

Cherry trees are very popular, but in wetter regions they are plagued by many disease problems. Smaller, more compact trees are easier to maintain. A 'Mazzard' rootstock will result in a tree that grows to 6 to 8 m (20 to 25 feet). A 'Colt' rootstock will give you a cherry tree of 4.5 to 6 m (15 to 20 feet) and the newer 'Gisela' grows to 3 to 4 m (10 to 13 feet).

If you get a self-fertile variety, many of which were developed in Canada, you won't need a second cherry to get fruit.

Sweet cherries (*Prunus avium* varieties)
- 'Angela'. Excellent midseason variety with large, black, crack-resistant fruit. Similar to 'Bing'.
- 'Bing'. The standard large, black fruit. Self-thinning. Delicious, very sweet fruit, but be warned that it cracks readily in rain. Midseason ripening.
- 'Compact Stella'. Early, large, black, oval-shaped fruit. Excellent flavour. Some crack resistance. A great pollinator for other trees. Semidwarf. Midseason ripening.
- 'Compact Van'. Black, shiny, firm, delicious fruit. Heavy-bearing tree. Great pollinator. Semidwarf. Bears early.
- 'Early Burlat'. Large, firm, meaty, delicious fruit. June ripening.
- 'Hardy Giant'. Very large, dark red fruit. Great pollinator.
- 'Lambert'. Large, heart-shaped black fruit. Excellent flavour. Cracks in wet weather.
- 'Lapins'. Long, dark red fruit. Very productive. Crack resistant, delicious fruit. Self-fertile.
- 'Rainier'. Large, 'Royal Anne' type yellow fruit with a red blush. Good pollinator. Midseason ripening.
- 'Royal Anne'. Light yellow, pink-blushed fruit. Great flavour. Usually ripens after wet weather in late July. Used for maraschino cherries.
- 'Sam'. Very large, jet-black, heart-shaped fruit. Good crack resistance. Good pollinator. Early ripening.

- 'Sunburst'. 'Stella' type. Self-fertile. Good crack resistance. Large, sweet black fruit and a heavy yield. Early ripening.
- 'Sweetheart'. Excellent, very late variety, with large, delicious fruit. Self-fertile.

Sour cherries (*Prunus cerasus* varieties)

All sour cherries are self-fertile.
- 'Montmorency'. Bright red, glossy fruit. Firm, juicy, commercial variety. Excellent for cooking. Late ripening.
- 'Northstar'. Bright red, tart fruit. Good for pies. Compact tree. Midseason ripening.
- 'Schatten Morello'. Deep crimson-black fruit. Firm with good flavour. Virus-free variety. Late ripening.

Crabapples (*Malus* varieties)

Crabapples are now available on dwarfing rootstocks, making them much more useful in the home garden. They are the single best pollinators for apple trees, often having a pollinating viability of over three weeks. They may be a little tart, but I enjoy them fresh, and as jelly and preserves.
- 'Dolgo'. Crisp red fruit. Juicy, acidic flavour.
- 'Hyslop'. Yellow, with crimson blush.
- 'Maypole'. A columnar tree with masses of beautiful, carmine blossoms for two weeks in early May. Attractive bronze foliage. Large, purple-red crabapples throughout summer. Matures in mid-September. Excellent for jelly.

Figs (*Ficus carica* varieties)

Figs need a sheltered location on a south-facing wall, out of cold winter winds, as they are zone 7 trees. They need a lot of water—never let them dry out—but they should be planted in well-drained soil. Try to stress the plants by limiting root growth. Virtually pest free, they will produce two crops a year, but only one will mature. There are green varieties and purple varieties. All are self-fertile.

Pruning is relatively easy—simply thin out weak and inward-growing branches, and prune fast-growing main stems back in June. To increase production, all young shoots should be pinched back so that only five leaves remain to encourage new fruiting shoots. Embryo figs will appear late in the season and will develop the following year.

Figs are ready to harvest when fully coloured fruit hangs heavily downward. The first crop matures in July; the second in late September where climate permits. Enjoy their exceptional fresh flavour or dry them for winter use.
- 'Brown Turkey'. Rich, sweet flavour. Red flesh. First crop ripens in July. Second crop does not mature in northern climates.
- 'Desert King'. Green fig. Very sweet and quite hardy.
- 'Mission'. Purple skin with strawberry flesh. Good fresh or dried.
- 'Peter's Honey'. Light yellow-green fruit with dark amber flesh. Excellent flavour.

Nectarines (*Prunus persicus* var. *nectarine*)

Nectarines are cousins of peaches. Their fruit is smooth-skinned and they have a slightly different flavour. They are not as self-fertile as peaches, but peaches will pollinate nectarines.
- 'Crimson Gold'. Medium size, with deep crimson skin and bright orange flesh. Smooth texture. Excellent flavour. Ripens early August.
- 'Early Sungrand'. Large variety with orange-red skin and yellow flesh. Semi-freestone. Excellent flavour. Very productive. Ripens mid-August.
- 'Red Gold'. Very large, red-gold fruit with yellow flesh. Freestone. Excellent flavour. Ripens early September.

Genetic dwarf nectarine
- 'Golden Prolific'. Grows to less than 2 m (6 feet). Medium-large, yellow, freestone fruit. Ripens late August.

Oriental pears (*Prunus pyrifolia*)

These are often dubbed "pear apples" and are perhaps the most distinctly flavoured pears. The flesh is crisp and aromatic. Very hardy trees (zone 4) are disease resistant and simple to grow. The fruit is generally small on young trees, but will increase in size with age. Thin out clusters of fruit in spring to improve size. Most trees are partially self-fertile, but will increase production with a pollinator.

Young trees are susceptible to bark canker in wet areas, and should be sprayed with copper three times during the dormant season.

- 'Chojura'. Attractive, russeted, round fruit. Sweet, aromatic flavour. Ripens in August.
- 'Hosui'. Large, russeted variety with juicy, fine textured flesh. Ripens in August.
- 'Kousi'. Attractive yellow fruit with exceptional flavour. Ripens late August.
- 'Shinseiki'. Yellow, sweet and flavourful. Ripens late August.
- '20th Century'. Attractive yellow skin with white, finely textured, juicy, sweet flesh. Stores well. Needs a pollinator. Ripens early September.

Peaches (*Prunus persica* varieties)

A fresh, ripe peach straight from the tree has to be one of the great joys of life. Peach trees are self-fertile, bear soft fruits and require warm summer temperatures or a west-facing wall to set fruit and ripen. They produce fruit as a young tree on lengthy one-year-old branches.

Peach leaf curl is a real problem in wet climates but can be prevented by growing them in greenhouses, under building eaves or by spraying with fixed copper three times over the winter. 'Frost' (zone 4) is leaf-curl resistant.

Standard trees grow to 7.5 m (25 feet), but can easily be kept shorter with pruning. Semi-dwarfs grow up to 5 m (16 feet). Bees are important for pollination. If your peach blooms before the bees are out, use a fine brush to pollinate the flowers.

- 'Early Red Haven'. Red-blushed fruit. Firm and delicious. Semi-clingstone. Early ripening.
- 'Fairhaven'. Medium-large, yellow-blushed fruit with good flavour. Freestone. Good for canning. Ripens mid- to late August.
- 'Frost'. Medium-sized, red-blushed fruit with good flavour. Freestone. The most leaf curl–resistant variety available.
- 'Glohaven'. Quite large, round, with minimal fuzz and red skin. Great flavour. Freestone. Ripens in early September.
- 'Pacific Gold'. Heavy-producing coastal variety, which is supposed to have leaf curl resistance. Ripens mid- to late August.
- 'Redhaven'. Medium to large fruit. Freestone. Bright red colour and excellent flavour. Good disease resistance. Ripens mid-August.
- 'Renton' ('Surrey Peach'). Freestone. Developed on the west coast to have good leaf curl resistance. Flavourful fresh or canned. Ripens mid- to late August.
- 'Rochester'. Large peach with firm, yellow flesh. One of the best for the west coast. Ripens mid-August.

Genetic dwarf peaches
- 'Bonanza'. Will grow to 2 m (6 feet), but produces fruit as a young tree. Yellow flesh. Freestone with bland flavour.
- 'Empress'. Grows to less than 2 m (6 feet), producing red-skinned, tasty fruit. Clingstone. Ripens early August.
- 'Golden Glory'. Grows to less than 2 m (6 feet), with large, yellow-fleshed fruit. Freestone. Delicious, juicy flavour. Ripens late August.

Pears (*Pyrus communis*)

Pears are closely related to apples, but pear trees generally live longer, need more sunshine than apples and are more sensitive to frost, both at blossom time and in winter. They are somewhat less prone to pests and diseases, but the fruits have a shorter storage life. Fire blight is one of the major problems in wet areas and choosing disease-resistant varieties is recommended. Like apples, most pears need a pollinator.

Nearly all pears are grown on 'Quince A' rootstock and grow to about 4.5 m (15 feet). They tolerate damp or heavy soils and do well in a wide variety of climates. Dwarf pyramids are available that grow to only 1.5 m (5 feet).

- 'Anjou'. Large green fruit, ripens to a greenish yellow. Light yellow flesh is sweet and juicy. Resists fire blight. Ripens mid-September to late October.
- 'Aurora'. Large, bright yellow fruit. White, sweet, juicy, aromatic flesh. Resistant to fire blight. Long keeper. Ripens in September.
- 'Bartlett'. Medium-sized green fruit, turns yellow when ripe. Flesh is firm and has good flavour and aroma. Heavy producer. Ripens mid-August. Good for canning.
- 'Beurre d'Hardy'. Short, plump, russeted fruit with distinct flavour. Leaves turn bright red in autumn. Pick in mid-September when fruit is hard. Vigorous tree.
- 'Bosc'. A great winter pear with russeted skin. Long-necked, juicy fruit with a spicy flavour. Stores well. Heavy producer. Ripens mid-September.
- 'Clapp's Favourite'. Flavourful yellow pear with a red blush. Very juicy and sweet. Heavy producer. Ripens mid-August.
- 'Comice'. Medium to large fruit with blushed, russeted skin. Flesh is firm, fine-textured and sweet. A great dessert pear. Ripens late August to September.
- 'Conference'. One of the most reliable producers under a variety of weather conditions. Long, narrow, partially russeted fruit. Flesh is firm. Very juicy with good flavour. Good keeper. Ripens late September.
- 'Eddie'. Sweet, brown-skinned pear with smooth flesh. Medium-sized fruit is great for canning. Ripens in August.
- 'Flemish Beauty'. Large yellow fruit with red blush. Flesh is tender and juicy with good flavour. Ripens early September.
- 'Highland'. Good-sized pear with yellow skin and some russeting. Yellowish flesh is juicy and sweet. Ripens late September.
- 'Red Anjou'. Large, oblong, smooth-skinned red fruit with yellowish flesh. Stores well. Ripens mid-September.
- 'Red Sensation'. Medium to large white-fleshed fruit is good fresh or canned. Ripens early September.
- 'Seckle'. Small, reddish, russeted fruit with delicious flavour. Generally not grown today, but it has been rated as one of the most durable pears. Ripens late September.
- 'Sierra'. Large, light green fruit with superb sweet flavour. Bears early and heavily. Stores well. Ripens late September.

est bets

Flavour
- 'Anjou'
- 'Aurora'
- 'Clapp's Favourite'
- 'Comice'
- 'Sierra'

For a wet climate
- 'Aurora'
- 'Bosc'
- 'Comice'
- 'Conference'
- 'Highland'

◀
Conical forms in containers offer a wonderful opportunity to use vines almost anywhere. Sometimes vines can be used in combination for interesting year-round effects.

▼ For a fast cover, especially over chainlink fences, try vines like the hardy Virginia creeper. Its fall colouring is spectacular.

▲ Magnolias create a grand display each spring even though the blossoms are somewhat short-lived. *Magnolia sargentiana robusta* is one of the earliest and largest-flowering varieties for zone 6 or warmer climates.

◀ Mophead hydrangeas not only look great in summer but their dried blossoms can add a nice touch to your Thanksgiving or Christmas decor.

▼ Variegated hydrangeas offer colourful foliage as well as flowers. There are many new variegated varieties, such as 'Lemon Wave'.

▶ A weeping blue spruce adds colour and an interesting form to the landscape.

▼ Espalier apples can be trained as a low fence around your garden.

◀ With proper watering and fertilization, 'Collonade' and dwarf fruit trees can be successfully grown in containers. Annuals make a nice accent.

▼ Bulbs are quite at home in containers and window boxes, but need winter protection in colder areas.

▶ Dwarf crabapples are great pollinators for apple trees and make fine jellies and preserves. Left on the tree, they also work well attracting birds.

▼ Woodland settings are ideal locations for naturalizing bulbs. Species bulbs both naturalize and multiply far better than hybrid varieties.

◀ Old-fashioned tiger lilies provide a real treat of midsummer colour. They also naturalize easily for a bigger and better show every year.

▼ Blending biennials like violas with bulbs, you can greatly enhance the effect with fewer bulbs.

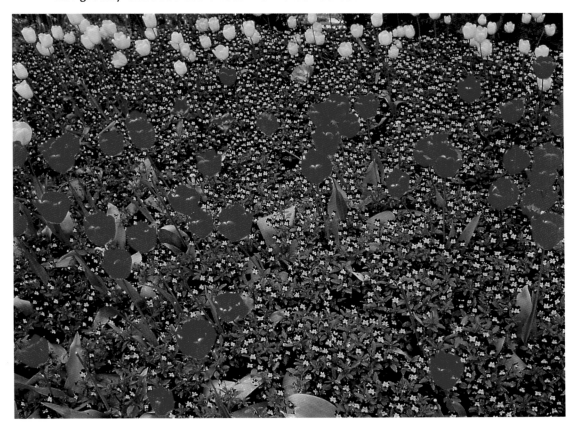

◀ We've used creeping jenny as a year-round groundcover at Minter Gardens (zone 6) with great results. I particularly enjoy the yellow flowers all summer long.

▼ Some easy-to-grow groundcovers can create bright splashes of colour in an otherwise plain landscape.

- 'Spartlett'. Very large green fruit turns yellow when ripe in late August. Excellent flavour. Often considered a big 'Bartlett'. Good fresh or canned.

Plums (*Prunus* varieties)

Plums are certainly one of the most popular stone fruits and relatively easy to grow in the home garden. They do get large, however, so they need some space, even if they are grown on the compacting 'St. Julien' rootstalk.

Two distinct families of edible plums are the European plum (*Prunus domestica*) and the Japanese plum (*Prunus salicina*). The European varieties are later blooming, making them more suitable for areas with late frosts.

There is a wide range of European plum sizes and colours, each with its own unique flavour. Prune plums are by far the most popular, not only because of their high sugar content, but also because of their versatility for use in jams, preserves and as dried fruits.

When it comes to flavour and size, however, it's hard to beat the Japanese varieties. They are so juicy, a napkin is standard equipment. These are the real dessert plums, with a wonderful taste.

All plum trees are vigorous and open vase pruning is the best way to train them in the home garden. I've found the Japanese varieties to be most vigorous, needing yearly pruning to keep them in check. European trees require somewhat less work as they mature.

Most varieties require a pollinator that blossoms at the same time, and that's why you need at least two Japanese or two European varieties near each other to cross-pollinate. Prune plums, 'Green Gage', Victoria', and 'Santa Rosa' are self-fertile varieties, making them ideal for gardens with no room for two plums.

Plums need an open, sunny area on high ground with well-drained soil to perform best.

European plums
- 'Blue Damson'. An old-time favourite. Small, round, blue-black plum with green flesh. Very sweet when ripe and excellent for preserves.
- 'Green Gage'. A midseason, distinct, green-yellow plum with amber flesh. Very sweet with excellent flavour. Ideal for canning and jam.
- Italian prune plum. There are many early and late varieties available, but I like the mid- to late variety for its flavour. Good for jam and dried fruit. It's self-fertile.
- Peach plum. Mid- to large-sized plum with yellow and bluish skin. Firm flesh is very sweet. Ripens easily in August.
- 'Yellow Egg'. Very large, oval to oblong fruit with a unique, sweet flavour that is not acidic. Ripens late August.

Japanese plums
- 'Redheart'. Very dark, dull red skin with heart-shaped fruit and bright, sweet red flesh. This medium-sized variety matures in late August.
- 'Santa Rosa'. Very large, round, red-purple fruit with outstanding flavour. Ripens mid-August.
- 'Satsuma'. A midseason, dark red fruit with solid, meaty, sweet flavour.
- 'Shiro'. Large, round, yellow fruit with a pink blush. Very juicy and sweet. Ripens late July.

Nut Trees

As a rule of thumb, nut trees are far easier to grow and less prone to disease than most fruit trees. You don't have to pick them either, because when they're ready to harvest, the nuts drop to the ground. It's as simple as that. Just be sure to get there before the hungry squirrels arrive.

A Nut Sampler

Almonds (*Prunus dulcis* varieties)

Almond trees are hardy as ornamentals, but need a warmer climate (zone 6) for nut production. Early frosts will freeze nut crops. They need warm summers, and on the west coast should be grown on a south- or west facing wall to prevent leaf curl. Trees grow up to 9 m (30 feet). Fruit has a leathery, flattened shell with the almond pit inside. Harvest when hulls split. Most varieties need pollination, but some are self-fertile.

- 'All In One'. Self-fertile. Large, sweet nuts. Ready late September to October.
- 'Hall' ('Hall's Hardy'). Good-sized nuts with hard shells. Partially self-fertile, but produces better with another variety.
- 'Ne Plus Ultra'. Large soft-shelled nuts. Excellent pollinator for 'Nonpareil'.
- 'Nonpareil'. One of the best varieties. Nuts are easily shelled.

Chestnuts (*Castanea* varieties)

Chestnuts are large trees that will grow up to 18 m (60 feet). Good production takes about 8 to 10 years. At first the nuts will be quite small, but they'll size up. They need another seedling variety or another variety for pollination.

- Chinese edible chestnut (*Castanea mollisima*). Large, fast-growing variety. Needs sunny location and warm summers to produce.
- 'Colossal'. A cross of Japanese and European chestnuts. Large, high-quality nuts.

Filberts (*Corylus maxima* varieties)

Filberts are very productive trees for the home garden. Zone 6. They grow up to 6 m (20 feet). Two different varieties are needed for pollination. Grafted trees with two varieties on one rootstock are now available, so only one tree is necessary.

Grow single-stem for a more attractive tree. You may have to share the nuts with the jays.

- 'Barcelona'. Large round nuts with good flavour. Vigorous and productive. Good pollinator.
- 'Butler'. High-yielding tree with very large, oval nuts. Good flavour.
- 'Daviana'. Large, long nut with good flavour. Pollinator for 'Barcelona'.
- 'Du Chilly'. Large, long, high-quality nut. Spreading habit. Nuts are slow to drop. Pollinator for 'Barcelona'.
- 'Nooksack'. Long, large nuts. Productive.
- 'Royal'. Large, thin-shelled nuts. Productive. Excellent flavour.

Walnuts (*Juglans* varieties)

Walnuts are large trees, up to 24 m (80 feet), that are late to leaf out. They are self-fertile but will produce more nuts if another variety is planted nearby.

Black walnut (*Juglans nigra*). Fast-growing shade tree with high timber value if grown in colder climates. Small, hard-shelled nuts are difficult to crack.

Other nuts

Heartnuts are medium-sized trees that grow to 9 m (30 feet) and have long, tropical-looking leaves. They produce tasty, heart-shaped nuts that hang in strings. Zone 4.

Another nut tree worth considering is the buartnut. It's a fast-growing, hardy cross of butternut and heartnut. Trees grow up to 15 m (50 feet). Nuts are large and flavourful. Zone 3.

Carpathian walnut (*Juglans regia*). Large, soft-shelled nut. Very cold hardy. Zone 4.
- 'Franquette'. Large, late-producing tree with medium- to large-sized, thin-shelled nuts.
- 'Manregion'. Large, thin-shelled, round nuts with delicious flavour. Never bitter.
- 'Russian'. Medium-large, soft-shelled nuts of excellent flavour. The hardiest walnut variety.

Common Fruit and Nut Tree problems

Apple scab
Scab is a fungal infection that primarily affects apples grown in wet climates. Both leaves and fruit can be affected; leaves become yellow and distorted before dropping. Brown-black spots appear on the fruit, which can be distorted as well.

The best way to prevent apple scab is to plant a scab-resistant variety, such as 'Liberty' or 'Freedom'. If you don't want to replace a diseased tree, you may be able to control the disease by following a strict spray program.

To control scab, remove affected leaves and twigs and destroy them. Spray on a regular basis with several fungicides, alternating among them so the scab doesn't become resistant to any one fungicide. Choose a time when it's not raining, or not expected to rain.

The first spraying should take place at bud break in the spring, when new green growth is 4 cm (1½ inches), again at the pink bud stage, a third time one week after the petals fall, and a fourth time two weeks later. You will have to respray as the season progresses.

Follow label directions carefully, noting times between application and harvest. Do not spray when flowers are in bloom. It might affect pollination or harm bees.

Canker or gumming
This bacterial disease hits cherry, apricot or peach trees, usually when they're young. Buds will look dead or weak and twigs will die. Often there is gumming early in the spring. Leaves may have small black spots and fruit can show many small, sunken black spots.

There is no cure for this disease and it will sometimes kill the tree over two or three seasons. You can control it by removing as many of the infected parts as possible. When you prune, make sure you cut at least 30 cm (1 foot) below the infected point. Disinfect pruning equipment between cuts by dipping blades in a sterilizing solution, such as one part bleach to nine parts water.

Spraying with a copper solution during the dormant season will control the canker in future years, but it won't cure it.

Fruit drop
Fruit drop can be caused by abrupt weather changes from very cool to hot when the blossoms are forming fruit. This causes stress and leads to fruit drop. The other cause is a lack of boron, which helps with fruit initiation.

Treat by watering the roots with a boron solution of 15 ml per 4 litres (1 tablespoon per gallon) of water. Don't overdo it, because too much boron can be harmful to plants.

Boron deficiency
If the meat of nut trees is shriveled, or the tree is not producing, it could be due to a boron deficiency. Boron can be found in fruit tree and vegetable fertilizers as a nutrient.

Small Fruits
Delicious strawberries, mouth-watering blueberries, tart raspberries and sweet grapes are some of the most rewarding plants for any home gardener. They are very easy and surprisingly inexpensive to grow. New and better

varieties are more compact, produce amazing crops in small spaces, and have even fewer disease problems than in the past. Once established, they can produce, with minimal maintenance, a steadily increasing crop for years to come.

Small fruits have become invaluable for today's gardens because many of them can play multipurpose roles. Compact-growing blueberries, such as 'North Blue', are fine ornamentals for fall and winter colour, and blend nicely with other plants in the landscape as well as providing fruit. Cranberries make superb groundcover in a variety of conditions. Grape vines can create shade and privacy screens if trained in clever ways. Everbearing strawberries look fabulous trailing over cement walls and produce a continuous supply of delicious fruit. Blueberries even work well as hedges.

Small, or soft, fruits is a general term for a diverse group of plants that bear soft-skinned, juicy fruit and grow on bushes, perennial plants or vines. As a rule, they require very little space, generally produce crops within a year or two of planting and are easy to prune and harvest. Soft fruits are generally divided into five categories.

- Herbaceous perennials, such as strawberries.
- Cane fruits, such as raspberries and tayberries.
- Bush fruits, such as currants, blueberries and gooseberries.
- Vines, such as grapes and kiwi.
- Broadleaf evergreen shrubs, such as lingonberry and cranberry.

Care and feeding

Soft fruits require attention to keep the plants looking good and producing an optimum amount of fruit. Yearly pruning, fertilizing and some disease and pest control are necessary to keep them in top form. After all, the ultimate goal is to grow the most delicious, fresh and pest-free fruits possible.

Late winter or early spring is the best time to purchase plants. Certified stock, inspected by the department of agriculture as pest and virus free, is essential, especially in strawberries and raspberries, to make sure you do not bring an infection into your garden. With so many disease problems, I would never accept strawberry or raspberry plants from a neighbour, just to be safe.

Bare-root plants are the usual fare first thing in the season; be sure you use the mud slurry method (see chapter 1, "Seeds and Seedlings") to get them off to a fast start. Container-grown plants afford the opportunity to plant any time during the growing season.

For the best results, plant small fruits in a sunny location. Blueberries, currants and gooseberries will tolerate partial shade. I like to suggest incorporating them into the landscape as part of the vegetable garden, as well as in the flower and shrub beds. Be creative and fit them into your overall scheme.

Small fruits perform best in deep, well-drained soil with a fair bit of organic matter. None of them like wet feet, especially blueberries and raspberries. An annual mulching of well-rotted manures will provide plenty of vigour, but be sure to counter that with the appropriate high-phosphorus and high-potash fertilizer to control rapid growth and produce large crops of good-sized fruits. In areas of heavy rainfall and heavy soils, an annual application of dolomite lime (along with a soil test, of course) is an excellent idea.

A Small Fruit Sampler

A great many varieties of small fruits are available, from strawberries and blueberries to currants and Saskatoon berries. Each performs well in certain conditions and hardiness zones. Following is a small selection.

Strawberries (*Fragaria* varieties)

Strawberries are the ideal fruit crop for gardeners with very limited space. The plants are low-growing perennials that fruit either in June or throughout the summer. June-bearing strawberries are the largest and most productive. They begin producing in June and finish in mid-July. Everbearing varieties produce delicious, but slightly smaller, berries from July until frost, and will produce the first season. Day-neutral varieties are newer on the market. They start producing earlier and keep on as late as the everbearing types. They produce medium-sized berries with excellent flavour. Day-neutral and everbearing varieties can be grown in containers and as trailers.

Plants can be grown in rows, beds or pots. Plant them 45 cm (18 inches) apart. They make a pleasant groundcover when not fruiting. Most of their root system is in the top 15 cm (6 inches) of soil, making them sensitive to an excess or a lack of moisture. Always buy certified, disease-free plants.

Strawberries perform best in sunny locations in soils enriched with organic matter. As with most small fruits, they won't tolerate wet feet. For this reason, raised beds are a great idea. This also helps restrict runaway plants from taking over. Set out plants in March or early April when soil begins to warm up. Keep new plants well watered.

June-bearing strawberries

Name	Zone	Description
'Hood'	3	Large, dark red berries on upright, vigorous plants. Great flavour.
'Rainier'	3	Large firm berries. Ripens late. Excellent flavour.
'Shuksan'	3	Large, glossy, wedge-shaped fruit. Productive, very hardy.
'Sumas'	3	Heavy yielding. Lighter red than 'Totem', and not as firm. Good flavour.
'Totem'	2	One of the most popular because of the large berries, productivity and hardiness.

Everbearing strawberries

Name	Zone	Description
'Fort Laramie'	2	Excellent vigour and runner production. Good for hanging baskets. Hardy. Good flavour.
'Quinault'	3	One of the best everbearing varieties. Moderate early crop. Heavy July and September crops. Large, good-quality fruit.

Day-neutral strawberries

Name	Zone	Description
'Fern'	3	Good flavour. Medium-sized fruit.
'Hecker'	3	Wedge-shaped fruit. Red throughout, with excellent flavour.
'Tristar'	2	Most popular day-neutral. Hardy, disease-free. Medium-sized fruit.

Cane fruits

These are some of the most delicious, mouth-watering berries you'll ever come across. Trained properly, they take up little space and are easy to maintain. Blackberries, loganberries and boysenberries are a real treat eaten fresh with ice cream. They also make great preserves and can't be beat for pies. Always select thornless varieties for ease of harvesting and pruning, unless you are a bit of a masochist.

Fruiting canes of most cane fruits should be cut back after fruiting to allow new shoots to develop for the coming season.

Try to harvest soft berries in the morning, when the sun has dried the dew. This will prevent heat buildup and lengthen their shelf life. Refrigerate immediately in shallow containers and they should last up to a week. Blackberries keep longer, and all of them freeze well for later use.

Blackberries (*Rubus fruticosus*)

Name	Zone	Description
'Black Satin'	4	Excellent, newer, heavy-yielding, semi-erect, thornless variety. Improved quality and hardiness.
'Chester'	4	Hardiest of all blackberries. Thornless. Heavy producer.
'Marion'	4	Midseason. Medium-large fruit. Very thorny. Good flavour and long producing.
'Sylvan'	4	New. Most highly recommended. Very vigorous and good disease resistance.
'Thornless Evergreen'	4	Top commercial variety in Oregon. Fruit is large and sweet. Plants are very productive.

Raspberries (*Rubus idaeus* varieties)

Although many gardeners still favour summer-bearing raspberries, there is a trend toward the everbearing varieties, which have a longer period of fruit production, usually from late August through October. Raspberries are terrific small fruits because they are very productive while being relatively carefree. Instead of planting rows of plants, the new concept is to plant clumps supported by a simple peonylike set of rings to keep them compact and prevent them from spilling all over the place. They require well-drained soil and a sunny location.

After the canes have borne fruit, cut the old canes to the ground over the winter, allowing the new canes to grow and produce fruit the following season. For everbearing raspberries, you can push fruit production to later in the season by cutting all of the canes to the ground in early winter.

To avoid losing flavour, try not to wash raspberries before eating. Instead, wipe them gently with a clean cloth. They do, of course, make great jam, and even better, a summer drink called raspberry vinegar.

Everbearing raspberries

Name	Zone	Description
'Amity'	4	Medium-sized, tasty fruit. Ripens mid-August. Good disease tolerance.
'Autumn Bliss'	4	Fabulous, fall-fruiting variety that has become the recommended choice. Significantly bigger yields than other varieties. Good flavour.
'Fallgold'	4	Unique yellow raspberry that fruits in mid-August. Good flavour and sweetness. Medium-sized berries.
'Heritage'	4	Large and firm. Heavy producer. Ripens September to late fall.
'Summit'	4	Small- and medium-sized fruit with great flavour. Very high yields and early producing.

Summer-bearing raspberries

Name	Zone	Description
'Blackcap'	3	Erect bushes. Small, black fruit. Big yields, high quality.
'Boyne'	2	The hardiest variety. Produces medium-sized, deep red, delicious fruit. Vigorous.
'Chilliwack'	4	Promising newer variety similar to 'Willamette'. Smooth canes resemble 'Skeena'. Good disease resistance.
'Meeker'	4	Large and firm. Vigorous and productive. Good on poorly drained soils.

Summer-bearing raspberries, cont.

Name	Zone	Description
'Qualicum'	4	Large, firm dark red fruit and almost thornless stems. Productive, with a long fruiting season. Ripens early.
'Skeena'	4	Firm, red fruit of excellent flavour, produced on sturdy, upright canes.
'Tulameen'	5	Exceptionally large fruit and good colour. Ideal for home garden due to fruit size and extended harvest season.

Other hybrid berries

Name	Zone	Description
Boysenberries	4	The thornless, evergreen boysenberry has large, purple fruits with a flavour akin to wild blackberry. Tart and juicy fruit can be used fresh or in preserves. Drought resistant.
Loganberries	5	Large red berries are very juicy and more tart than a boysenberry. Good for jams and preserves. Thornless stems.
Tayberries		This is a hybrid of blackberry and red raspberry. Similar to loganberry, but larger and sweeter. Matures a deep purple colour. Slightly thorny plants. Berries are good fresh, canned or frozen.
Youngberries	4	A light purple hybrid loganberry cross.

Bush fruits
Blueberries (*Vaccinium* varieties)

Blueberries prefer moist but well-drained, slightly acidic soils. They fall into three categories: lowbush, which grow 30 to 60 cm (1 to 2 feet) high, half-highbush, which grow 1 to 1.2 m (3 to 4 feet) and highbush, which reach 1.2 to 1.8 m (4 to 6 feet). They need little care or pruning and are generally disease free. Prune out damaged or dead branches.

Blueberries are great ornamentals with flowers, fruit and brilliant fall foliage. They are self-fertile, but produce far more with two or more different varieties. Try early, midseason and late varieties to extend your harvest period as long as possible.

Blueberries

Name	Zone	Description
'Blue Crop'	4	Midseason highbush. Top commercial variety. Good flavour and one of the easiest to grow. Large berries.
'Brunswick'	3	Low-growing and spreading blueberry from New Brunswick. Unique as a groundcover with outstanding wild blueberry flavour.

Name	Zone	Description
'Darrow'	5	Late highbush. One of the largest-fruiting blueberries ever developed. Produces 2.5-cm (1-inch) fruits that are slightly flat and delightfully tart.
'Dixi'	4	Late highbush. Large, sweet berries. Old-time favourite.
'Earliblue'	5	Very early highbush. Medium to large berries.
'Herbert'	4	Late highbush. Very large, tasty berries.
'Jersey'	4	Late highbush. One of the oldest varieties. Medium-sized fruits. Very sweet.
'Legacy'	5	Evergreen highbush. Late. Noted for great flavour and productivity. Ideal ornamental.
'Northblue'	3	Midseason highbush. Compact, to 90 cm (3 feet). Medium-sized berries. Very sweet with good colour. Extremely productive.
'North Country'	3	Early midseason, half-highbush. Compact, to 60 cm (2 feet). Closest to wild blueberry flavour. Medium-sized fruit. Great foliage for use in landscapes.
'Northland'	3	Early to midseason highbush. Medium-sized berries with excellent flavour.
'Northsky'	3	Midseason lowbush. One of the most compact, to 45 cm (1½ feet), and cold hardy. Nice-looking plants. Medium-sized fruit with wild flavour.
'Patriot'	3	Early highbush. Large berries with good flavour. Highly productive. Will tolerate some wet soils.
'Rancocas'	4	Midseason highbush. Small and medium-sized fruits. Very prolific. Good flavour.
'Sunshine Blue'	5	New midseason. Semidwarf evergreen. Hot pink, showy flowers. Produces large crop of medium-sized, delicious berries with tangy flavour. Great ornamental.
'Top Hat'	3	Midseason lowbush. Small- and medium-sized berries on novel-looking plants. Ideal for pots or as bonsai specimen.
'Toro'	4	Midseason highbush. Giant, delicious, sky-blue berries. Great ornamental with flowers that turn from hot pink to bright white. Fall foliage is one of the most brilliant.

Currants (*Ribes* species)

Red and white currants can be eaten fresh or used for pies, jams, jellies or wine. Black currants are best used in those outstanding jams and jellies. Most fruit is produced on last year's stems, so new growth must be allowed to develop each year.

Do your pruning any time between fall

dormancy and late winter bud swelling. With black currants, once established, cut out some old stems, leaving a good number of last year's stems to produce a new crop. Red currants are pruned to leave an open centre. Fruit is carried on stubby side shoots off the main branches. It is often difficult to tell which stems are which in winter, so be sure to mark the new stems after pruning with a dab of white latex paint.

If you live in areas of late spring frosts, the 'Ben' group of black currants blooms later than the rest of the currants.

Red currants (*Ribes silvestre* varieties)

Name	Zone	Description
'Red Cherry'	3	Mildew resistant. Heavy producer.
'Red Lake'	3	Large red fruits. Good flavour.
'Red Start'	3	Bright fruits. Heavy yields. Tart.

Black currants (*Ribes nigrum* varieties)

Name	Zone	Description
'Ben Lomond'	3	Late flowering. Heavy yields. Sharp flavour. Good mildew resistance.
'Ben Sarek'	3	Dwarf plant, 1 to 1.2 m (3 to 4 feet). Mildew resistant. Pick fruit as it ripens.
'Boskoop Giant'	3	High-yielding variety with medium-sized, deep black fruit.
'Consort'	3	Resistant to pine blister rust. Very productive. Musky flavour.
'Crandall'	3	Attractive yellow flowers; a great ornamental.

White currants (*Ribes silvestre* varieties)

Name	Zone	Description
'White Versaille'	3	Somewhat of a grape flavour.
'White Grape'	3	Long trusses of pale, white fruits with excellent flavour. Ripens mid-July.

Elderberries (*Sambucus canadensis* varieties)

American elderberries are fast-growing shrubs, producing clusters of red berries that turn black in July and August. The berries aren't the best for fresh eating, but are ideal for preserves, pies, jams and wine. Get two different varieties for most effective pollination.

Name	Zone	Description
'Johns'	3	Large white flowers and juicy, purple berries.
'Variegata'	4	A newer plant that's a great novelty. Fine-looking ornamental that's ideal for any garden.
'York'	3	Produces large, white flowers, followed by giant, juicy, purple berries.

Gooseberries (*Ribes uva-crispa* var. *reclinatum*)

Ripe gooseberries are simply outstanding and well suited to fresh eating, pies and preserves. The plants prefer partial shade and tend to do best if pruned to an open centre form, like red and white currants. As with currants, when gooseberries are mature, keep removing old canes, allowing the growth of new, more productive canes.

Name	Zone	Description
'Captivator'	3	Semithornless. Nice to work with.
'Hinnonmaeki Red'	2	Large, red sweet variety from Finland. One of the best on the market. Mildew resistant.
'Hinnonmaeki Yellow'	2	Large, green sweet variety from Finland. Mildew resistant.
'Oregon Champion'	3	Large green-yellow fruit. Sweet when ripe. Mildew resistant.

Highbush cranberries (*Viburnum trilobum*)

Highbush cranberry is a large shrub, growing to 2.4 m (8 feet). Clusters of white flowers produce red berries in August. The berries are used for wines, syrups and preserves.

Saskatoon berries (*Amelanchier alnifolia*)

Saskatoon berry is a fast-growing shrub that can reach 2.4 m (8 feet). It has white flowers in May, followed by purple berries in August. Brilliant orange-red autumn foliage.

Vines

Grapes (*Vitis vinifera*)

Grapes are a favourite fruiting vine, producing fresh table grapes, juice and wine. They take about three years to begin producing from a two-year old plant, and will continue to be productive for about 40 years. Training and pruning are essential if you want to produce quality fruit and an attractive vine.

They need as much sun as possible. South- or west-facing walls are ideal to absorb extra heat for maturing some of the late varieties. Grapes prefer a well-drained, porous soil. Plant in early spring, three to four weeks before the last frost. After setting out plants, remove all but one cane with up to six buds. Maintain an adequate water supply, at least until mid-June. Supplement rainfall to ensure the equivalent of 2.5 cm (1 inch) of rain per week. Keep vines off the ground, tied to either stakes or trellises.

There are all kinds of pruning and training methods for grapes, depending upon whether they are American or European types, but for the home gardener, the "four-cane Kniffin method" is perhaps best. This simply means training one main stem to produce four side shoots (two on each side) growing along a double set of wires. Grapes fruit on second-year, finger-thick wood, so prune out weak canes, leaving 10 to 15 buds per cane for fruit production on a mature vine. This should give you 40 to 60 buds on a well-developed framework.

Prune in late February to minimize winter damage. Don't worry about heavy sap flow in late winter; it will not weaken the canes. Prune in summer to remove rampant growth, allowing more air and sunlight in and around the grapes to prevent mildew.

Here are some of the best bets for grapes, divided into seedless, table and wine grapes.

Eating grapes

Name	Zone	Description	Uses
'Arkansas Blue'	4	Large, blue seedless. Wonderful flavour.	Table
'Bath'	4	Medium-sized black grapes. High quality.	Wine, fresh
'Buffalo'	4	Medium-large fruit. Compact bunches.	Table, wine
'Canadice'	4	Early red seedless. Ripens two weeks before 'Concord'. Disease resistant.	Table
'Concord'	4	Late. Medium-large, sweet grapes. Most widely grown variety in North America.	Table, jelly, juice
'Flame'	4	Red seedless. Crunchy. Good flavour. Needs south-facing wall. Premium greenhouse variety.	Raisins, table
'Fredonia'	4	Huge black grape, similar to Concord. Good for arbours. Best juice grape.	Table, juice
'Glenora Seedless'	4	Excellent table grape. Medium-sized, seedless, black.	Table
'Himrod'	5	Early white seedless hybrid. Medium-sized and sweet.	Table

Name	Zone	Description	Uses
'Homestead Red'	5	One of the finest red seedless. Medium-large fruits. Excellent flavour.	Wine, table, dessert
'Interlaken Seedless'	5	Medium-sized seedless amber grape. Sweet.	Table
'Lakemont'	4	White seedless. Excellent sweet flavour. Tight clusters.	Table
'Niagara'	4	Large, white grape. Sweet, tangy flavour. Top eating variety.	Table
'Simone'	4	Blue seedless. Heavy producer. Good flavour. Needs south-facing wall.	Table
'Sovereign Coronation'	4	Newer blue seedless. Nice fruity taste. Ripens late September.	Table
'Thompson Seedless'	4	Medium, greenish white to golden seedless grapes. One of California's leading shipping grapes.	Table, wine

Wine grapes

Name	Zone	Description	Uses
'Cabernet Sauvignon'	4	Black Bordeaux with claret flavour.	Wine
'Golden Muscat'	4	Large, high-quality, green yellow grapes.	Table, wine
'Pinot Noir'	4	Excellent red grape. Early, good quality.	Wine
'Riesling'	5	Medium, white grape of high quality. Excellent wine.	Wine
'Seibel 9549'	4	Black European hybrid. Important commercial wine variety.	Wine
'Wonder'	4	Large, juicy, seeded blue grape. Disease resistant. Produces large crops of sweet, tangy grapes. Large clusters, red juice.	Fresh, wine or juice

Kiwis (*Actinidia* varieties)

Kiwi is a great garden plant that's soared in popularity in recent years. With attractive, pest-free vines, kiwis are valued as a tasty fruit loaded with vitamins. The vines will often reach 10 m (33 feet) or more. They need well-drained soil and summer irrigation. There are two species: hardy kiwis (zone 3) and fuzzy kiwis (zone 6). They will generally pollinate each other, but it is essential to have male flowers blooming at the same time as the females. Male and female flowers should be reasonably close for good pollination. The hardy kiwi 'Arctic Beauty' blooms too early to pollinate other species. Most varieties require three years of proper pruning before they will bear fruit.

The main objective in pruning kiwi vines is to establish and maintain a well-formed framework of leaders and fruiting arms. Problems arise mainly because of the exceptional vigour of the vines, which must be controlled by both summer and winter pruning. Fruit normally develops on the first three to six buds of the current season's growth. Although new canes may arise at almost any point on the vines, usually only the canes from the previous season's growth bear fruit.

In summer, cut back unwanted new lateral canes and maintain stringent control of spur growth along the permanent fruiting arms. Frequently cut back the spur growth hard and keep the spurs as short as possible.

Winter pruning is needed to remove fruiting wood which is beginning to lose vigour or has become overcrowded, or to encourage new growth. Start pruning in December after the crop has been harvested and when the vines are dormant. If you leave it until early spring, sap exudes freely from the cuts and may cause considerable dieback of valuable wood. Generally, try to finish winter pruning by about mid-January.

Right from the outset, concentrate on developing well-defined, permanent leaders, adequately furnished with an evenly spaced system of fruiting arms. This is the most important measure to take for easy, quick management of the vines over a period of years. Usually about one-third of the fruiting arms need to be removed from the vines each year. Cut them back hard to the permanent leaders. Growth from older wood after removal of the spent fruiting arms does not normally bear fruit in its first season. In the second year, however, these new arms produce several fruiting side laterals, and in the third year further fruiting laterals develop.

Usually some winter pruning is needed to control them. Shorten these laterals to two buds beyond where the previous season's fruit was borne. One or both of these buds will grow in spring and produce fruiting wood. Two buds should be left in case any injury prevents or destroys growth from the terminal bud in spring. This will allow the other bud to grow and bear fruit.

As a general rule, remove the fruiting arms after the third year. It may be a good idea to retain a fruiting arm with its system of laterals for a further season, but this will depend on the growth available for furnishing the vine. Keep the current fruit-bearing wood distributed as evenly as possible over the whole vine.

Hardy kiwis (*Actinidia arguta* varieties)

Name	Zone	Description
Actinidia arguta	2	Smooth-skinned, grape-sized delicious fruit. Needs male and female to produce fruit. Thin occasionally to control rampant growth. In dormant season, prune to one or two trunks. Cut out excess branches. Fruit is borne on shoots one year or older.
'Issai'	4	Self-fertile variety that produces fruit on very young plants. Not as hardy as *A. arguta* and somewhat irregular in fruiting. Delicious, smooth-skinned fruit.

Fuzzy kiwis (*Actinidia deliciosa* varieties)

Name	Zone	Description
'Hayward'	6	Most popular female variety. Large, sweet, attractive fruit ripens in late October.
'Saanichton'	6	Selected on Vancouver Island, B.C. This variety will withstand cooler temperatures. Fruit is large and somewhat blocky. Ripens one week before 'Hayward'.
'Tomuri'	6	One of the best male pollinators. Prune into compact form to avoid wasting space. All you need is a few male flowers to pollinate.

Hops (*Humulus lupulus*)

Hop vines are familiar to those who like to brew their own beer. The twining vines are hardy to zone 5.

Name	Description	Uses
'Cascade'	Excellent, all-purpose strain.	North American beers
'Hallertauer'	Pleasant aroma.	German lagers
'Mount Hood'	Medium bitterness. Mild flavour and good aroma.	Bavarian-style lagers
'Nugget'	A strong, bitter hop.	Stouts and ales
'Willamette'	Spicy aroma.	North American beers

Broadleaf evergreens

Cranberries (*Vaccinium oxycoccus*)

One of the forgotten small fruits, the cranberry has a lot of potential for the home garden. It thrives in boggy soils, but will perform well in almost any garden situation as long as it has moist, peaty soil. It grows fine in containers and is a great ornamental. Cranberry also makes a low evergreen groundcover, with tiny leaves and wiry stems.

The tiny, whitish pink flowers produced in May are quite beautiful, but the huge berries are even nicer looking. Just don't eat them raw—sauce them along with lots and lots of sugar. Zone 4.

Other broadleaf evergreens

These dwarf, hardy, broadleaf evergreens grow up to 40 cm (16 inches) in height and make excellent groundcovers. They bloom in spring and again in summer, and all bear small berries. They need a well-drained, acidic soil but require little in the way of nutrients.

Creeping blueberry (*Vaccinium crassifolium* 'Well's Delight'). Zone 5. Evergreen groundcover 13 to 20 cm (5 to 8 inches) tall. Leaves are similar to Japanese holly. New spring growth is brilliant bronze. Small white flowers and tiny, dark-coloured berries are ideal for wildlife. Plant at 60-cm (2-foot) intervals for use as groundcover.

Lingonberry (*Vaccinium vitis-idaea* var. *minus*). Zone 2. Native to eastern Canada and the U.S. Pink blooms cover 10- to 13-cm (4- to 5-inch) plants that grow in mounds. Pea-sized fruits. Ideal rock garden plant.

Mountain cranberry (*Vaccinium vitis-idaea* 'Koralle'). Zone 3. European favourite, growing 30 to 35 cm (12 to 14 inches). Heavy yields of bright, pea-sized fruit. Popular commercial variety.

12 Bulbs

About 300 years ago, when Ahmed III ruled the Ottoman Empire, tulips were more precious than gold. In fact, this period in the early 18th century is often referred to as the "age of tulips." Many books were written on the subject, and a surviving catalogue from the era lists 890 varieties for sale. In one recorded transaction, a single tulip was sold for two loads of wheat, four loads of rye, four fat oxen, eight fat pigs, twelve fat sheep, two hogsheads of wine, four barrels of beer, two barrels of butter, a thousand pounds of cheese, a bed, a suit of clothes and a silver beaker. That's one expensive bulb!

Today tulips aren't quite so pricey. They have been classified into 15 groups, with hundreds of recognized varieties under cultivation. Tulips remain popular, and they have a place in our gardens, but there are many other bulbs that deserve our attention.

The word "bulb" is often used loosely to refer to a broad group of plants. A true bulb is composed of layers of compressed immature leaves, called scales, attached to a basal plate which is actually a squashed stem. A corm is the swollen base of a stem with stored nutrients. A tuber also stores food as a fleshy outgrowth of an underground stem, but unlike bulbs and corms, it has no basal plate. Tuberous roots are modified roots that store food underground. Rhizomes are thickened stems that grow horizontally, producing roots below and sending up shoots from their upper surface. It's important to distinguish among them to be botanically correct, but for our purposes, they're lumped together under the umbrella term of bulb.

Planting and Care of Bulbs

Bulbs only keep a certain length of time on the shelf in the garden store before they start to deteriorate. Buy them as close to planting season as possible, and inspect the bulbs before you buy. Reject any that are rock hard, soft or with signs of rot on them. If you're storing them, don't keep them in the garage where they can be exposed to vehicle exhaust fumes—even small amounts of carbon monoxide will damage them.

Planting

Each bulb has its own needs in terms of planting, but as a rule of thumb, planting depth should be two and a half to three times their width. A bulb that's 2.5 cm (1 inch) across should be planted about 8 cm (3 inches) underground. Tiny bulbs, like snowdrops, should be planted about 8 cm (3 inches) deep to provide them with a little extra winter protection in colder areas (zone 5 or below). In cold climates, plant all bulbs deeper in order to keep them below the frost line.

Most bulbs recommended for fall planting are hardy to zone 3. Bulbs planted in spring for summer and fall flowers are generally more tender, and must be lifted out of the ground and stored for the winter. (See "Storing bulbs" near the end of this chapter.)

In warmer climates, it's always safer to plant bulbs on the shallow side, rather than too deeply. Foxtail lilies will rot if planted too deep. These unusual octopuslike roots produce beautiful, tall, yellow flower spikes. Plant them shallowly (15 cm/6 inches deep), mulch them well for winter and, come spring, they will have miraculously pulled themselves down to the depth they want.

Special treatment

While most bulbs are easy to plant—with the pointy side up—some types need special handling. Anemones, for example, should be soaked in warm water to soften them up for planting. An hour or two should do it, but I've left some in a bucket overnight and planted them the next day. Hyacinth bulbs, especially white ones, can cause an itchy sensation and a rash, so be cautious when handling them. Wear gloves, and don't scratch your skin.

All bulbs need well-drained soil so they don't rot in wet winters. Add sand to the planting hole to improve drainage and encourage a vigorous root system to develop. Sandy soil helps cure the bulb after it has bloomed by allowing the water to drain away, leaving it in good shape for the next growing season. Starter fertilizers, such as bone meal (0-15-3) or liquid root booster, are valuable in encouraging quick rooting when planting new bulbs.

In Holland, as a guest of the International Flower Bulb Institute, I learned that virtually all their bulbs grown commercially were planted in raised furrows in sandy soil and mulched lightly with straw. The soil was kept moist through irrigation, even during cool spring weather. These are the best bulb growers in the world, so you won't go wrong following their lead when it comes to looking after bulbs in your garden.

Plant bulbs in groupings of threes, fives and sevens, and concentrate on those that naturalize and multiply so there is an ever-increasing sea of flowers in your garden. Plan how to combine bulbs with other plants to create the effect you want. For instance, you may want a splash of colour all at one time as a focal point, or you may choose a succession of flowers over the course of the year.

Feeding

What you do this year determines how successful you'll be next year, so feed bulbs in the critical period after they have finished blooming. I use either a simple 6-8-6 fertilizer or a bit of fish fertilizer once the flowers begin to decline. They also respond well to nutrients in the form of well-rotted manures and good garden compost. Once they finish flowering, they have a very narrow window of time in which to build up their flower buds for the next year.

I've been asked whether using bone meal on bulbs at planting time is grossly overrated.

It's not: bone meal is high in phosphorus, which helps in the rooting process, and it is organic. It will help stimulate growth if you are late planting your bulbs, although it only kicks in once the roots start to grow. A bulb is a perfect entity unto itself and it doesn't need any help to bloom.

Year-round considerations

Many people complain about the "before and after" problems of bulbs. Beforehand, you are waiting for the plant to develop, then there's the payoff of a relatively short burst of colour as the plant flowers, and finally you have to wait a seemingly interminable length of time while the foliage dies down, building up the plant's energy for next year—so-called "ugly time." You can avoid some of these problems by careful bulb selection. I prefer using species of bulbs, rather than fancy hybrids, because they generally tend to be more weather tolerant, less subject to insect and disease problems, and have nicer looking foliage after flowering. They also give you a bonus in the ground by rapidly multiplying each season for bigger and better displays in years to come.

One of the biggest mistakes people make when it comes to bulbs is pulling them out of the ground or cutting the leaves off as soon as they finish flowering. After they flower, bulbs build up energy to form next year's flower. The leaves manufacture nutrients that are carried down to the bulb. Always allow the leaves to die down naturally before removing the bulbs or cutting the leaves off. If you can't stand the sight of them, there is a solution. Plant the bulbs in 10-litre (2-gallon) containers and pop the pots into the ground, covering them right up. When the flowers finish, pull out the pots and place them out of sight until the leaves die down naturally. Don't forget to water and feed them, though, so the bulbs can rebuild for next year.

Companion planting

The other solution is to cover up the later stages of the bulb's season with companion planting. Pros always companion plant, and you should too. By using complementary or contrasting plants you can heighten the effect of a display, and use fewer bulbs in the process. Many perennials and biennials bloom earlier and longer than the bulbs to give continuous colour in a given bed. The best benefit is the masking effect of these plants. They cover up bulb foliage as it dies down. In late spring you can set out the annuals to continue the parade of colour a gardener loves to see.

If you're using biennials as companion plants, plant them by early October, while there is still some growing weather left, and they can become established before the winter frosts. Start them from seed in June, or buy plants in 10-cm (4-inch) pots in the fall. Violas and pansies are the number-one choice in this category. There have been a lot of improvements in the last few years, and you might want to replace older ones with the more compact, long-blooming, weather-tolerant types.

Winter warm spells

Sometimes there is a spell of early warm weather that encourages some bulbs to shoot up out of the soil long before their normal blooming time. If this happens in January or February, another blast of winter is likely.

Don't worry too much about frost damaging the plants, because the flower stem is still inside the bulb. You might see some burn on the leaves, though. If this is keeping you up at night, cover them with a thick layer of bark mulch to protect them until the real spring arrives.

Small- to medium-sized flowering varieties handle the cold weather better, and research has proven that small-flowered violas are hardy to zone 4.

Miniature violas, or johnny-jump-ups, also make terrific companion plants. 'Helen Mount' (cream, yellow and purple), 'King Henry' (purple) and 'Baby Franjo' (yellow) explode with colour winter and spring.

Another option is English daisies. The big ones, such as those in the Super Enorma Strain, are especially showy. The reds and pinks are effective, but I prefer the whites, with their fluffy 6-cm (2½-inch) flowers, which can bloom from January through to June, and set off any bulb to great effect. The smaller-flowering button types are even more weather tolerant.

In shady areas, forget-me-nots are too often overlooked as companion plants. You can find them in shades of blue, white and pink. They're easy to grow and will flower until the end of June.

Planting under trees

Remember that you don't have to restrict your bulbs to the flower beds! Grow them in your lawn and around the base of trees for a very pleasant springtime effect. There's nothing better than the sight of flowers surrounding a lush tree.

Achieving this can be difficult, however, if you have a thick canopy overhead. Thin out the tree and limb up a few branches if possible to allow more light in. Some bulbs work better than others in this situation. Try smaller bulbs, such as snowdrops, winter aconites, snow crocus and squills.

Plant daylilies with daffodils: the old daffodil leaves are easily concealed by the similar foliage of the daylilies. Tulips and bearded iris make a similar foliage blend, and the beautiful, big leaves of hostas and ferns develop later in the season and are ideal for masking many spent bulbs. Experiment and observe what's happening in your garden to come up with the right combination. Above all, be adventurous with your bulb selection and search out the more exotic varieties.

Timing is the key

There are two rules to remember when planting bulbs.
1. Plant in fall for spring colour.
2. Plant in spring for late summer and fall colour.

Timing is the secret. Know approximately when bulbs flower in order to achieve the colour display you're looking for, especially when planting with a mix of flowering perennials and shrubs. Most people want a steady flow of colour through the growing season, rather than a single eruption of colour where you blink and it's all gone until next year.

Keep in mind that nature doesn't work on the same calendar every year—a very late spring will delay flowering up to three weeks, or an early spring can put the blooming schedule on the fast track. But with a little planning and a smattering of luck, you can time your floral displays. For example, a terrific early-blooming rhododendron is 'Praecox', and its lavender flowers look great with Siberian squills or 'Cream Beauty' crocus. Underplant February daphne with *Crocus vernus* 'Remembrance'. Beautybush (*Kolkwitzia amabilis*) is a great match for the blue spikes of common camass.

Storing bulbs

Some bulbs are hardy and will survive winter in the ground, but others need to be lifted out of the soil and stored. Some nonhardy varieties will survive a mild winter, but you risk losing them if the frost is too deep.

Store bulbs in a cool but not freezing spot. Most heated basements are too warm (except for gladioli and begonias, which don't like really cool temperatures). A garage with reasonable air circulation is an ideal location, but don't expose bulbs to a lot of carbon monoxide, which is generated when you have a car warming up in an enclosed area.

A certain amount of humidity is also required to successfully store bulbs and most of our homes are too dry. Seek out that special, perfect spot and remember where it is!

Let all bulbs dry out completely before storage. Ideally, place them in wooden boxes packed with vermiculite or very dry peat moss. Vermiculite gives extra protection by absorbing excess moisture.

Bulbs as perennials (naturalized bulbs)

Inexpensive, easy and beautiful are the three words I've come to associate with naturalized bulbs. These no-fuss flowers are especially important if you don't have a lot of time to fiddle about. They have remarkable staying power and will put on a display that will be really something to look at in years to come.

The trick is to plant them in locations where they can be left alone to naturalize. This means that once they're planted in fall, they're left on their own to spread and regenerate each year. Don't plant naturalizing varieties in your annual beds where they will be accidentally dug up, watered when they should be resting, and looking unsightly when you want to pop in early colour. Try using them under small ornamental trees like Japanese maples or weeping flowering cherries. Flowering shrubs like white forsythia, buttercup winterhazel and winter jasmine wonderfully complement early-flowering bulbs.

Miniature daffodils like 'February Gold' or 'Hawera', as well as the taller trumpet varieties 'Carlton' and 'Golden Harvest', make ideal underplantings for red-flowering currants, yellow-flowering Oregon grape, the purple-leafed smoke tree and the coppery new growth of photinia. The dark-foliaged roses are a natural for shared space with yellow daffodils.

Hostas, daylilies, ornamental grasses and evergreen perennials like coralbells, foamflower and lungwort not only complement early-flowering bulbs but cover up the foliage after they bloom. And don't forget your lawn as a place to plant bulbs. Early-flowering crocus, snowdrops and winter aconites are good choices. It will take about three years for the bulbs to really become established and put on a great show for you. Remember to allow the leaves to die down naturally before you lop them off with the lawn mower.

Most naturalized bulbs can be left in the ground, but lift them every three years to check for disease and insect damage. At this time you can divide them to increase your stock.

Extending the tulip season

Why be part of the pack and have your tulips bloom in April along with everyone else? I plant a few of the Fosteriana hybrid tulips, some of the earliest flowering ones I have found, in the warmest south-facing spot in my garden. With shelter from the winter winds and extra heat from the early spring sun, March-flowering tulips will bloom in late February, a welcome splash of colour, and a surprise to anyone who visits.

On the other end of the scale, plant a late-blooming variety, such as 'Princess Margaret Rose' in a cool, shady location where it will bloom in late June.

Best bets—spring-blooming naturalizing bulbs

Spring bulbs that come up each year in greater numbers are something I look forward to every year. Here are some favourites to blend with your late winter and early spring plants. For more suggestions, see "Best bets—spring-blooming, shade-loving bulbs."

Bulb name	Height	Approx. flowering time	Comments
Dutch crocus (*Crocus vernus*)	10-13 cm (4-5 in.)	March-April	Naturalize easily in lawns. Plant them in groups of 5 to 10 and in a variety of colours for best effect. Come in yellow, white, purple and striped.
English bluebell (*Hyacinthoides non-scripta,* also sold as *Scilla non-scripta* and *Scilla nutans*)	25-30 cm (10-12 in.)	May-June	Stems loaded with deep purple bells are truly a magnificent sight. Prefer light shade but will tolerate sun.
Glory-of-the-snow (*Chionodoxa luciliae*)	15-20 cm (6-8 in.)	March-April	Fast becoming a favourite spring performer. Pale blue, starlike flowers with white centres line 15- to 20-cm (6- to 8-inch) stalks.
Lily leek (*Allium moly*)	15-20 cm (6-8 in.)	June	One of the last fall-planted bulbs to bloom. Bears masses of yellow, starlike flowers. Naturalize easily in well-drained soil.
Narcissus 'Tête-à Tête'	20-25 cm (8-10 in.)	April	Multiflowered stems produce numerous small yellow trumpets. They tolerate wind and rain and because of their tiny leaves, they don't look messy after flowering.
Species tulips	15-25 cm (6-10 in.)	March-May, depending on variety	Single or multiple flower stems open up in the sun, then close at night. Well suited to rockeries.
Windflower (*Anemone blanda*)	8-10 cm (3-4 in.)	March-May	These underrated, low-growing, daisylike flowers are outstanding in a spring garden. They naturalize easily and the white, pink and blue tones combine well with early red tulips.

Best bets—spring-blooming, shade-loving bulbs

We all have a shady spot or two in our gardens that needs a lift of colour. Some sun-loving bulbs perform well in shade, while others prefer to be in the shadows all the time. Here are some great bulbs for those shady spots, and don't forget to blend them with hostas, corydalis and other shade-loving perennials for combinations that will add something special to your garden. All are naturalizing.

If you're choosing one of the lilies, remember that it's the Oriental lilies that have that wonderful perfume. Both *Cyclamen coum* and *C. hederifolium* need to be planted under trees with excellent drainage.

Bulb name	Height	Approx. flowering time (zone 4)	Colour
Cyclamen (*Cyclamen coum* and *C. hederifolium*)	10-15 cm (4-6 in.)	December to March (*C. coum*); August to November (*C. hederifolium*)	pink, red and white; pink, mauve and white
Dogtooth violet (*Erythronium dens-canis*)	15-20 cm (6-8 in.)	May	yellow
Grape hyacinth (*Muscari armeniacum*)	15-20 cm (6-8 in.)	April-May	blue, white
Lilies, most varieties (*Lilium*)	1-1.5 m (3-5 ft.)	June-October	various
Persian fritillaria (*Fritillaria persica*)	60-90 cm (2-3 ft.)	April	purple bells
Snowdrop (*Galanthus nivalis*)	10-13 cm (4-5 in.)	January-March	white
Winter aconite (*Eranthus hyemalis*)	8-10 cm (3-4 in.)	January-March	buttercup yellow

Best bets—spring-flowering, fragrant bulbs

Thinking back, I can still remember the wonderful perfume of the late-blooming 'Cheerfulness' narcissus in my grandfather's fields as a child. It's something I'll never forget, and a reminder that while colour is great, fragrance is a heavenly bonus. Always plant fragrant bulbs on the lee side of the wind, and near windows and frequently used doorways to take full advantage of the perfume as it wafts through your garden.

Bulb name	Height	Approx. flowering time (zone 4)	Colour
Common hyacinth (*Hyacinthus orientalis* hybrids)	25-30 cm (10-12 in.)	April	white, blue, pink, yellow and red

Bulb name	Height	Approx. flowering time (zone 4)	Colour
Grape hyacinth (*Muscari botryoides* 'Album')	15 cm (6 in.)	April	white
Madonna lily (*Lilium candidum*)	60-75 cm (24-30 in.)	June and July	white
Narcissus 'Bridal Crown'	45-50 cm (18-20 in.)	April	double cream
Narcissus 'Yellow Cheerfulness'	40-45 cm (16-18 in.)	May	double yellow
Netted iris (*Iris danfordiae*)	10 cm (4 in.)	February-March	yellow
Oriental lily hybrids (*Lilium* varieties)	.5-1.5 m (20 in.-5 ft.)	June-October	pink/white
Single early tulip (*Tulipa* 'General de Wet')	38-45 cm (15-18 in.)	April	orange
Spring snowflake (*Leucojum vernum*)	20-25 cm (8-10 in.)	February-March	white

Best bets—fall-blooming

Not many people plant bulbs in the summer, but it's worth it. Every year I pester people to plant autumn crocus. Sometimes it's hard to find, and it's only available in the spring for a short time, but you can usually find saffron crocus. These are both great for fall colour, and reproduce well in well-drained, sandy loam.

Bulb name	Height	Approx. flowering time (zone 4)	Comments
Autumn crocus (*Colchicum autumnale*)	15 cm (6 in.)	September-October	Huge bulbs often produce 15 to 20 flowers in pink and lavender tones. The white variety 'Alba' naturalizes nicely under trees.
Banded crocus (*Crocus kotschyana*, also called *Crocus zonatus*)	8 cm (3 in.)	August-September	Rose-lilac flowers are abundant if given well-drained location with year-round moisture.
Saffron crocus (*Crocus sativus*)	25 cm (10 in.)	September-October	Lilac-purple blooms have distinctive yellow stamens. It takes 2500 stamens to make 31 g (a single ounce) of saffron.
Showy autumn crocus (*Crocus speciosus*)	13-15 cm (5-6 in.)	September-October	The earliest of all fall-flowering crocuses. Attractive, deep violet-blue flowers. Needs year-round moisture.

Bulb name	Height	Approx. flowering time (zone 4)	Comments
Winter daffodil (*Sternbergia lutea*)	15 cm (6 in.)	September–October	Golden yellow crocuslike flowers. Needs well-drained soil, warmth and sunshine to do well. Leave undisturbed to let it naturalize.

Forcing Bulbs for Indoor Use

Forcing bulbs means tricking them into flowering earlier than normal so you can enjoy them indoors from late December through March. It's not hard to do and there's a lot of satisfaction in watching the plant emerge and the flowers brighten up those dreary late winter months.

Forcing sounds like a taskmaster's threat, but really it's a wonderful opportunity to bring colour to your home a whole lot earlier in the year. It's like cheating winter just a bit, and when you smell the glorious perfume of hyacinths in January, it certainly makes a cold, grey winter a whole lot more bearable.

Garden centres often sell kits, complete with pots and a growing medium, for forcing bulbs. These prepared bulbs often include amaryllis, hyacinths, paperwhites or 'Grand Soleil d'Or' narcissus. You can use other bulbs that are not pretreated or prepared, if you follow these steps.

Best bets—bulbs for forcing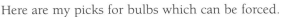

Here are my picks for bulbs which can be forced.

Bulb name	Colour	Weeks to bloom from fall planting
Common hyacinths (*Hyacinthus orientalis*)	yellow, blue, pink, white or red	8-10
Dutch crocus (*Crocus vernus*)	yellow, blue, purple-striped, white	8-10
Glory-of-the-snow (*Chionodoxa luciliae*)	blue with white centres	8-10
Grape hyacinths (*Muscari armeniacum*)	blue	8-10
Narcissus 'Tête à Tête', 'February Gold' or 'Bridal Crown' (very fragrant)	white/yellow	10-12
Netted iris (*Iris reticulata*)	blue	8-10
Siberian squills (*Scilla siberica*)	blue	8-10
Snowdrops (*Galanthus nivalis*)	white	8-10
Tulips—Double Early, Single Early or Triumph groups	various	8-12

1. Fill a 12.5- to 15-cm (5- or 6-inch) pot full of sterilized, well-drained potting soil and tap it down so the pot is almost full. A 12.5-cm (5-inch) pot will normally hold three bulbs, a 15-cm (6-inch) pot, five bulbs.
2. Plunge the bulbs into the soil so they're completely covered. Remember to keep the pointed side up.
3. Place a handful of heavy, wet sand on top of the soil and tap that down again. With lighter soils, which are basically peat moss mixes, the bulbs will just push themselves out. The sand will keep them in place.
4. Water thoroughly.
5. Put the pots in a cool spot. They must have a minimum of 8 to 12 weeks of rooting time at approximately 0° to 11°C (32° to 52°F). On the coast, you can simply put them outside in your garden and cover them with a thick layer of sawdust or bark mulch (make sure it's not cedar). This will keep the frost from penetrating. In cold areas, the ideal location is a cool shed that maintains these temperatures in the winter.
6. Water the pots. If the pots are outside, they will get plenty of rain. If you're doing this inside, keep the soil evenly moist.
7. Some varieties, such as prepared paperwhites, grow really quickly—in about 2 or 3 weeks. Others will take 8 to 9 weeks before they start coming up (see the chart below). As soon as the little flower stem comes up into the leaves, just above the bulb, bring the plant inside. Forcing produces a short, compact plant that looks very attractive in flower.
8. Once you bring them indoors, keep them well watered and remember that they like to be kept on the cool side. Lower temperatures mean longer life for the flowers.
9. Forcing bulbs kind of poops them out, so it's not a good idea to use the same bulbs for forcing twice in a row. Toss them in the compost after flowering—or plant them out in the garden in March and see what happens next year. But don't hold your breath!

Insects, Rodents and Rot

Bulbs face a number of obstacles before they bloom. Squirrels might ruin them by nibbling on them or simply dig them up, insects could attack them, or rot could set in and turn the bulbs to mush. Fortunately, bulbs are pretty tough and can beat these setbacks, especially if you plant them with a little forethought and good soil preparation.

One of the best ways of beating rot and insect infestations alike is to plant bulbs in very sandy soil. Many insects simply stay away and the sand keeps the bulbs from sitting in water all winter. If the bulbs do become infested with insects, you have a couple of choices. One is to identify the culprit and go after it. The other is to wait for the foliage to die down, lift the bulbs, inspect and discard the ones attacked by insects and move the rest to a new location.

"I planted my bulbs, but they never came up. What went wrong?" That's a common question, and it's often tied to a combination of soil conditions and location. Other than planting too deep, rot and fungus are the main causes of bulbs failing to survive. If they've spent the winter underwater, chances are they won't do very well come spring. This is remedied by working on your soil and improving drainage—something that will benefit most plants in your garden.

The other thing I recommend as a preventive is dusting bulbs with garden sulphur before planting; it's a great fungicide. Wettable sulphur powder is also good for keeping those pesky rodents at bay. Mix the solution in a pail, place your bulbs in a burlap sack and dip the whole thing in the sulphur solution. Not only will it deter rodents, it will also keep insects away. It's well worth the effort involved.

Lawns and Other Groundcovers

A lawn has universal appeal because of the green space it provides. It shows off other plantings to great effect, it's a wonderful place to play or relax, and—let's face it—it's hard to think of a better carpet to walk across barefoot. Parks, schoolyards, golf courses and ball parks wouldn't be much without lawns.

There are lots of ways to describe lawns, but in the end they are just another form of groundcover. In spite of some criticism about the ecological costs of lawns, they do have their good points. An average-sized lawn produces enough oxygen to keep a family of five breathing easier, stops soil erosion, helps fight pollution and is one of the easiest groundcovers to maintain.

They don't have to be a burden on the environment in terms of excess chemical use and the production of pollutants from gas-powered lawn mowers; lawns can be kept looking attractive with organic fertilizers or regular applications of slow-release nitrogen. Yearly aeration, applications of sand to keep the soil percolating, good mowing techniques and some weeding, watering, and overseeding can keep any lawn in great shape. Many lawn mowers are noisy polluters, but electric and hand mowers are good options.

Today, some folks are moving away from the traditional lawn to a combination of lawn and other groundcovers, such as ornamental grasses, shrubs and decorative paving. Smaller lot sizes and creative groundcoverings mean smaller and more effective lawn areas.

Lawns

Discussions about lawns go hand in hand with the topic of weeds and how to control them, and there will be more on that later in this chapter. But the overall message here is how to build and maintain a better lawn. Proper soil preparation, the right grass seed mix for your location and good lawn practices are the combination that will lead to a healthy, low-effort lawn for your home.

Maintaining established lawns

Aerating
If you have an already established lawn, give it the pencil test to see if the soil is sandy and loose enough to allow good air circulation and water drainage. You should be able to take a pencil and shove it right into the ground with-

out a lot of effort. If the pencil doesn't go in easily, it's time to rehabilitate that lawn with an application of coarse sand.

One of the best things that we can do for our lawns is to punch holes in the turf. Heavy winter rains and snow compact the soil, but aeration counters this, opening it up so oxygen can get to the roots. After aerating, broadcast a 1-cm (3/8-inch) layer of coarse sand over the lawn. It will make its way into the holes. This is an essential step that will help open up the soil, improve drainage and encourage the roots to grow deeper for a healthier lawn.

For the ultimate lawn, aerate spring and fall. Hand-held aerators are available, but for larger areas you will want to rent a mechanical aerator. Proper aerators remove a plug of soil, opening the roots to air and nutrients. The soil plugs can either be left on the lawn or raked up and placed in the compost.

You may think you can accomplish exactly the same thing by driving a digging fork into the ground, but the tines can actually compact the soil. There are special aerator shoes on the market that look like golf shoes on steroids. They punch holes into the soil as you walk over the grass. They're more of a novelty item than a serious gardening tool.

I can't emphasize enough how important aeration and sand are in making your lawn the envy of the neighbourhood. And the bonus is that this will cut down the need for chemical life support. Add the sand every year for at least three years to create good drainage.

Fertilizing

Your lawn needs a combination of things to grow, including well-drained soil, adequate moisture and proper mowing. If you have these three things in place, fertilizer provides the fourth key to a good lawn. Fertilizer improves the colour of the lawn and keeps it green for more of the growing season. Three fertilizer applications a year should give it good colour, promote vigorous growth and keep moss and weeds on the defensive.

The very best fertilizer for lawns, and reasonably environmentally friendly, is controlled-release nitrogen applied in spring, summer and early fall. The "controlled" part means the fertilizer has been polymer coated so that 25 to 40 percent of the nitrogen is released right away and the balance enters the soil over the following six or eight weeks. The nitrates are absorbed and utilized by the grass roots, the lawn maintains good colour and extra mowing is not necessary. The addition of iron or other micronutrients is often beneficial as well.

Fertilizers with formulations such as 16-4-4 or 18-6-12 are just fine as long as they contain slow-release nitrogen. I've had success with 16-4-4 with S.C.U., 18-6-12 with S.C.U. and 24-4-1 with S.C.U. (S.C.U. is sulphur-coated urea, which helps the nitrogen release slowly.)

To get brown grass toned up quickly in the spring, try using sulphate of ammonia 21-0-0, but slow-release nitrogen fertilizers are more environmentally friendly.

If you want to follow the organic route, you can try recycled waste fertilizers, such as canola-based fertilizer. They work slowly at first, but once the organic levels are up there,

---

Oops!

I once told a caller to my radio show to apply a 1-cm (3/8-inch) layer of sand to his lawn. He called back a couple of months later complaining it had smothered his grass. It turns out he didn't hear my instructions clearly and had covered his lawn with 8 to 20 cm (3 to 8 inches) of sand!

they maintain good colour for a long period of time. Organic lawn fertilizer costs more than nonorganics, and it takes up to a month before you see the results, but the benefits are spread over a longer period of time.

Always water in any fertilizer thoroughly after application, or apply it just before a good rainfall. Avoid this advice and you will likely find yourself the proud owner of a burnt lawn.

Applying fertilizer. A lawn spreader can be friend or foe, so take the time to use it correctly and avoid burning your lawn. For small lawns, simple hand spreaders are great. Larger lawns require a wheeled spreader that either broadcasts or drops the fertilizer as it goes.

Plastic or polyurethane spreaders are better than metal ones because they don't corrode as easily. Fertilizers will corrode and cake on with even a small amount of moisture, so always make sure you thoroughly wash out your spreader after use.

Keep your spreader in good shape by oiling the wheels and all moving parts. Make sure the hopper opens and closes easily before you put the fertilizer in.

To calculate the amount of fertilizer you need, measure your lawn (length × width = square feet or metres). When you get to the store, check the package for coverage and calculate the amount you need. Let's say you have 650 square metres (7000 square feet) of lawn. If one bag covers 370 square metres (4000 square feet), you will need two bags and have a bit left over at the end of the day.

Lime and lawns

Lime is another key to a good lawn in wet coastal regions, especially in heavy soils.

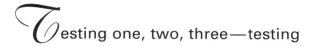esting one, two, three—testing

All the spreaders and the fertilizers are different, so don't trust the gauge on the fertilizer spreader. Don't. Don't. DON'T. (I really mean this.) Follow these steps instead.

1. Measure off a 3- by 3-m (10- by 10-foot) area on your lawn. This area, 9 square metres (100 square feet), will be your test patch.
2. Math test. How much fertilizer will we need for 9 square metres (100 square feet)? (Quick, phone your old math teacher.) Most slow-release nitrogen fertilizers come in 9-kg (20-pound) bags and cover 370 to 465 square metres (4000 to 5000 square feet). Divide the total area covered by the bag by the area of your test patch; it will be approximately 40.

This means you will need one-fortieth of the bag to cover your test patch—225 g of a 9-kg bag (half a pound of a 20-pound bag).
3. Once you've done your own math and know how much you need, get out the scale and put that much fertilizer in the spreader.
4. Set the gauge on your spreader.
5. Cover the test area with fertilizer. If you run out too soon or have some left over, you know the gauge isn't correct. Adjust accordingly and try another test patch.
6. When you have a correct reading, get an indelible felt pen and mark the correct setting for this brand of fertilizer on your spreader so you don't have to do this next time!

Winter rains drop the soil's pH and turn it acidic, which hinders grass growth and encourages moss to take hold. Lime raises the pH to counter these problems. Away from high rainfall areas, acid soil is not as much of a problem. Inexpensive pH testing kits will give you quite an accurate reading to determine whether you need to lime.

There are several types of lime. (See chapter 1, "Soil pH," for more information on lime.)

- **Dolomite lime** contains magnesium. Apply it in the fall for best results. It is slow-releasing and long-lasting. It's great for vegetable gardens as well.
- **Rapid lime** contains magnesium and is a faster-releasing lime. It's ideal if you forgot to lime in the fall and need quicker action. Hydrated lime is messy and difficult to apply, but it works very quickly to adjust pH.
- **Dolopril lime** is specially formulated for easier spreading. It provides fast release and long-lasting results.

Watering

Water very thoroughly once a week. Use a good sprinkler and don't even think about standing out there with a garden hose and sprayer unless you're willing to spend a few hours on the job. Water extra heavily where there are large trees that will suck moisture away from the lawn. Don't worry if you get some browning during the hottest part of the summer.

The most efficient time to water is early in the morning as the sun rises. Plants utilize water best at this time. If you water when it's

Tips for tip-top fertilizer application

- Plan out how you want to cover the area so you don't get confused.
- Keep moving at a steady pace. No running to catch the last inning of the ball game!
- When you stop, close the hopper right away, and mark the spot where you stopped.
- Always keep some fertilizer in the hopper. Spreaders don't work as well when they're too empty or too full.
- Don't let the fertilizer get wet, and make sure the grass is dry before you fertilize. Fertilizer burns wet grass.
- If you don't think it's coming out at the appropriate rate, dump the fertilizer back into the bag and check the opening of the hopper. Chances are some granules are caught in between the sliding parts and it is stuck open or closed.
- Keep away from trees and shrubs—make sure it doesn't spread too far if it's a broadcast type of spreader. Never use a broadcast spreader when applying fertilizer-herbicide combinations. It's too difficult to control the distribution pattern.
- Don't go over the same area two or three times. It's better to apply a bit less than called for rather than too much.
- Water in fertilizer immediately after application. This prevents burning and activates the fertilizer so it can release its nutrients into the soil.

hot, you lose water to evaporation before it penetrates the soil. Watering in the evening can contribute to disease problems.

Overseeding and scarifying

If your lawn is looking patchy, overseeding—applying new grass seed—is the answer. Almost every lawn can benefit from a yearly treatment in the spring, after you have done your weed or moss control. Where winters are harsh, some types of grass simply die out and need to be replaced in the spring. Overseeding is also a great way to help control weeds. When the weather warms up, overseed with a perennial rye grass. 'Elka II' is a dwarf perennial variety I've recommended for years but there are many other good varieties. Perennial rye grasses are usually fine-bladed and tremendously resilient, and they add new life and vigour to any lawn. There are specific grass seed mixes for different situations, such as high traffic or shade.

- Wait until daytime temperatures are in the 10° to 13°C (50° to 55°F) range.
- Give the seed a place to take hold by scratching up the lawn with a rake.
- Sprinkle a mix of moistened peat moss and sand into the target area.
- For a quick start-up, use 11-22-4 fertilizer.
- Apply the seed. The recommended rate for overseeding is the same as seeding a new lawn: 2.2 kg (5 pounds) of seed for every 93 square metres (1000 square feet) of lawn. However, I've had success with half that amount using the guidelines above.
- Perennial rye grasses should germinate in five to seven days. The seed needs a steady supply of moisture, so water daily for at least seven days if it doesn't rain.
- Don't mow until the new grasses take hold.

Overseeding reinvigorates old lawns and keeps weeds to a minimum. If you have some leftover seed, don't worry; lawn seed will last up to seven years if kept in a cool, dark place.

Mowing

How you mow your grass can make a big difference to the appearance of your lawn. Mowing in the same direction every time trains grass to lie flat and it will eventually form a matted mess, called thatch. Once it's lying down, you end up cutting the stems, which will cause brown patches on your lawn every time you mow.

To find out how good a job your mower is doing, get down on your hands and knees and take a gander at your grass. You should have short stems standing more or less straight up. If the grass is lying down on its side and the clippings are matting up, dethatch until the existing grasses are short and standing upright.

You can rent dethatching machines, which are best, or purchase a "roto-rake" bar for your rotary power mower. This bar has springs that go down into the thatch layer and tear it out. Simply replace the grass-cutting blade with the roto-rake bar and go to it. This is hard on power mowers, however. For a small lawn, a thatch rake and some effort should do the job.

To rake or not to rake?

If you don't allow the grass to get too long between cuttings and your mower chops the grass into fine pieces or you have a mulching mower, chances are it's okay to leave the grass cuttings on the lawn. This allows the nutrients to go back into the lawn.

Mulching mowers work great in dry climates, but in wetter areas there can be problems if wet clippings sit on the lawn. Rake if it's damp, and leave finely mulched clippings on the lawn when the weather is dry.

Some people make power raking an annual tradition, but it's not necessary unless your grass is lying down and matted. If you are going to power rake, do it in April when the grass is just starting to grow (that's also the time to aerate).

To prevent thatch from forming, mow your lawn low and never cut it in the same direction twice in a row. Keep it up all season and thatch won't have a chance to form. For most lawns, I recommend cutting the grass to about 5 cm (2 inches). This will also keep the grass growing straight.

Rehabilitating a tired old lawn

If you're fed up with a sad-looking lawn or have recently inherited one, here are the steps to get it back in top shape.
1. Check the pH of the soil with readily available pH test kits. It should be in the 5 to 8 range. Correct with lime or sulphate fertilizers if needed.
2. Aerate the lawn. Add a 1-cm (3/8-inch) layer of sand to invigorate the root system.
3. To achieve a fast green-up, fertilize with 21-0-0 and water immediately after putting it on.
4. If there are large bare patches, spread on a layer of moistened peat moss and sand and overseed the lawn with perennial rye grass. The peat moss helps hold moisture to aid germination.
5. Keep the lawn good and wet during this time. A week of warm weather in late spring will dry out the seed.
6. Apply more sand and aerate spring and fall every year for the next few years.
7. Enjoy your well-drained, moss-free, weed-resistant lawn!

Putting in a new lawn

Whether you're moving into a new house or changing the landscaping, here are the basic steps to follow when installing a new lawn. This groundwork applies whether you're growing from seed or bringing sod in from another location. The key is establishing good drainage and a suitable bed to get things started.
1. Use a safe herbicide over the entire area. Check with a licensed herbicide dispenser

Rolling out the lumps and valleys

Over time the surface of your lawn can sink in places, so that it seems as if you have a mini-version of the foothills. This is not only hard on your mower, it can shake your teeth right out of your head and leave you with a scalped lawn.

Small depressions can be filled with sand and finely screened soil, but rolling should be considered if there's more than one or two small spots to deal with. Lawn rollers are available for rent at garden centres. The best time to roll your lawn is just at the end of winter, when the ground is soft but not waterlogged. Hand rollers don't really do the job effectively unless you can find a good-sized one. Make sure you take your vitamins, because it can be quite a workout to roll a lawn. I recommend hiring someone with a small commercial roller if it's more than a small patch. Don't use too large a roller for fear of soil compaction.

at your local garden centre for the correct product that will kill all existing weeds and grasses, but will not harm the soil or leave any residue. Wait two to three weeks before moving on to the next step.

2. Work the soil to a depth of 15 to 20 cm (6 to 8 inches) with a heavy-duty rototiller.

3. Sift the earth to remove rocks, pebbles and debris. If you don't have enough topsoil, bring in a load of sand or light screened topsoil. It is very important to mix it in with the existing soils.

4. Roll it so it's firm with a heavy roller available for rent from garden centres or building supply centres. Go over the area in three different directions. Make sure it's as level as possible. It's much easier to level now than after the lawn is established.

5. Put a layer of peat moss on top to hold the moisture next to the seed.

6. Use a starter fertilizer with controlled-release nitrogen to get things going. One 10-kg (22-pound) bag covers 180 square metres (2000 square feet). If you're using sod rather than seed, you can lay it at this point.

7. Using a spreader, spread the seed in a north/south direction, then repeat in an east/west direction for a thorough distribution. Use approximately .5 kilogram per 20 square metres (1 pound per 200 square feet).

8. Keep it moist. Don't let it dry out for at least 7 to 10 days. This means in hot weather you may have to water two to three times per day until the grass is rooted.

9. Mow as soon as the grass is high enough to reach with a mower. Mowing is like pruning—it encourages the grass to fill in more quickly, leaving little room for those nasty weeds. Mow low initially, about 4 to 5 cm (1 1/2 to 2 inches), to prune young grass seedlings and make them fill out.

10. As with any lawn, the more you mow in different directions, the more full and weed free it will be.

Grass mixes

When selecting a lawn grass, the main considerations are the growing conditions in your area, the quality of your soil and how you'll be using the lawn. Advice from qualified local gardeners is always helpful. We often hear that some single variety of grass seed, such as one of the newer perennial rye grasses, is the solution to all grass problems. But remember, a blend of grass seed will create the best lawn. Some grasses, such as Kentucky blue grasses, perform best in hot, dry weather, while fescues tend to perform better in cool, damper seasons. With a blend, you can get good performance and a good look all through the year, except of course in areas with long winter cold spells.

Much like a carpet, you also have to consider how much traffic your grass will be dealing with. Will it be a play area, a place where people are always running about, or will you and the lawn mower be the only ones treading on your lawn? Some blends are made for heavy traffic, while others are best for show areas. Shady areas require a special blend as well.

The other consideration is the quality of the seed mix itself—I can tell you they vary greatly. Under government regulations, mixes

triped lawns

Some folks want to know how to get a lawn with those attractive stripes, just like you see in the professional ball parks. The answer is a reel mower with a roller, which flattens the grass in opposite directions as you mow, creating a striped effect.

labeled as number-one can contain a certain percentage of weed seeds, along with lower-end grass seed. Always purchase seed from a supplier who understands lawn mixes and can give you the percentage of each grass type and the percentage of weed seeds.

You can enjoy a lawn for a lifetime, so make sure it is the right blend for your needs. Following are blends for specific conditions.

Show lawn
- 40 to 50 percent top-quality perennial rye grass
- 20 to 25 percent top-quality blue grass
- 30 to 35 percent top-quality fescue grass

All-purpose back lawn
- 10 to 20 percent good blue grass
- 50 to 60 percent good creeping red fescue
- 25 to 30 percent good perennial rye grass

Shady area lawn
- 25 to 60 percent good chewing fescue
- 20 to 45 percent perennial rye grass
- 40 to 50 percent creeping red fescue

Low-maintenance lawn
- 40 to 50 percent perennial rye grass
- 20 to 30 percent hard fescue
- 40 to 60 percent creeping red fescue

Turf
Turf farms generally grow a range of grass blends for various situations, and for more instant results turf is certainly the way to go. It can be laid out both earlier and later in the season than lawn seed, providing an advantage for folks who require an instant green space.

The same soil preparation is required for turf as for seed (see "Putting in a new lawn," above). Once the soil is prepared and fertilized, the turf can be laid. It is very easy to lay out the turf rolls; just be sure to go in straight lines, and start the second row halfway up the first strip to achieve a staggered look. Roll it lightly to settle it in and water thoroughly to remoisturize and activate the roots. Try not to leave rolled turf in a stack for more than one day because it may dry out and turn yellow. It takes about a week to 10 days for turf to settle in.

 year-round lawn routine

 Spring. Aerate, apply sand, overseed, fertilize, weed.

Summer. Water deeply once a week, keep weeds in check, don't worry if it gets a bit dry or patchy in hot weather. Fertilize, but not during periods of extended drought without water being available.

 Fall. Apply dolomite lime to your lawn; keep cutting the grass and remove leaves as they fall. Aerate and apply sand. Fertilize.

 Winter. Keep debris off the lawn and be careful when applying ice-melting salt near grassy areas.

Lawn problems and solutions

Weeds and your lawn

The subject of lawn and weeds is becoming more controversial all the time. A couple of decades ago, it was simple—weeds were bad; grass was good. But the environmental movement has wisely questioned the overuse of herbicides because of environmental and health concerns.

The pursuit of the perfect patch of grass has now gone full circle to the point where people are actually encouraging wild weeds to grow in their front yards. This might irritate neighbours intent on a more manicured look for their property, but who's to say which approach is the right one? It's a battle that mixes concerns over the use of herbicides with neighbourhood aesthetics. In the end, it's up to you to come up with an acceptable strategy regarding weeds and your lawn.

To eliminate weeds, you need to get at their roots. The most effective way is to pull them out, roots and all, using a special weed trowel. They come in short- or long-handled versions. There are also some organic weed killers on the market. They work, but usually require more persistence than the nonorganic herbicides.

If you do choose a route that involves the use of chemicals, please do it wisely. Read the directions and remember, using twice as much as called for on the label doesn't mean it will work twice as well. Mixtures that "weed and feed" at the same time seem convenient but they often fail to do the best job. I prefer to make weeding and feeding two separate tasks. It's more work, but the results are superior.

The best defense against weeds in your lawn is thick, lush grass so weeds can't get a toehold and compete. Frequent mowing also helps by nicking weeds before they get a chance to flower. Sometimes, however, the weeds get ahead of us and some weed control is necessary. There are all types of weeds out there, and for the most part selective herbicides are the best bet. These kill the target weed, but don't harm lawn grasses. Some brand names will control up to 50 different types of weeds if applied correctly. As a rule of thumb, liquid weed killers are far more effective than granules because the liquid actually covers the weed foliage, whereas the granules are more of a hit and miss proposition. Read the labels and consult with the licensed herbicide dispenser at your local garden centre before buying.

Regardless of the method you use to eliminate weeds from your lawn, always follow by overseeding the lawn afterwards to keep the weeds from coming back. (See "Overseeding and scarifying," earlier in this chapter, for more information.)

Rules for applying weed killers

If you do use a herbicide, make sure to follow these guidelines.

- Choose a calm day, with little or no wind and temperatures in the 10° to 25°C (50° to 75°F) range.
- I've found that hose-end sprayers make the best use of time and product. Pressure sprayers and watering cans are the other options.
- Mix strictly according to directions. More is *not* better.
- Spray the weeds until they are wet, but not saturated. The spray should just be on the verge of dripping from the leaves.
- Stay well away from your neighbour's fence line and your garden and bedding plants. Keep checking the wind and make sure the spray is not drifting. If it is, stop and do it on another day.
- Do not water for 48 hours, and hope that the folks down at the weather office were right when they said "zero percent chance of precipitation."

- In two or three days the weeds should begin to curl their toes and die off.
- If some of the well-established weeds survive, a second application in 10 to 14 days should do the trick.
- Six days after applying the herbicide you can begin watering the lawn again to get the grass growing and fill in those dead patches.

If you have a real tough weed that normal broad-spectrum herbicides will not control, then you have to use something that will take it out without harming the soil. There are some very selective herbicides for specific weeds. Ask at your local garden centre, where qualified and licensed personnel will be able to make the proper recommendations.

Very tenacious weeds, such as couch grass and velvet grass, require a different approach. The best solution I've found is to use a non-selective, non-soil-harming herbicide as a spot weed killer. You have to be careful with it because it will kill desirable lawn grasses as well as the weeds.

One method of localizing the effect of spot weed killers is to cut a hole in the centre of a piece of cardboard. Place the target weed in the centre of the hole, so that the cardboard protects the lawn or other areas you don't want doused with herbicide.

Moss control

Moss in the lawn is a common complaint in wet areas. The quick solution is to apply iron sulphate, but you can only do this in good weather. You need temperatures above 10°C (50°F) and no rain for 48 hours after application. Iron sulphate, sold under various brand names and labeled as a "moss killer," comes in both liquid and granular forms.

Moisten the moss first and then apply the moss killer according to the directions on the package. Once the moss blackens, rake it out and dispose of it in the trash. But don't stop there—take steps to prevent the moss from coming back.

Poor drainage, acid soil and a lack of nitrogen are the main factors that allow moss to pop up in our lawns. With a little effort in the spring and fall you can develop a long-term approach to keeping moss under control in your lawn.

- Drain water away from heavy or low-lying soils. Take the two-prong approach of aerating and adding a 1-cm ($3/8$-inch) layer of sand every year. It will do wonders for the drainage as it works its way down into the soil.
- Keep the pH of the soil high by treating it in the fall with dolomite lime.
- Keep nutrient levels high so your grass is healthy and can overcome the moss.
- Control moss spores everywhere else in your garden, such as on trees, your roof or walkways.
- Increase light where possible by pruning shade trees.
- Wait about a week after you remove moss and then overseed the lawn. (See "Overseeding and scarifying," earlier in this chapter, for more information.)

Mushrooms

Organic matter in your lawn encourages mushrooms to grow. It could be from compost, including mushroom and cow manures, that you've put on the lawn (one reason I don't recommend using manures on lawns). Spores can also blow in on the wind and take hold given the right conditions.

Try to get rid of mushrooms as soon as they appear by raking them off. This will prevent them from reproducing. Lime the lawn, aerate, add 1 cm ($3/8$ inch) of sand over top, and this may help. There is no easy solution—just keep removing mushrooms as they appear throughout the season.

Fairy ring

This is a fungus that forms in a ring. It is usually prevalent where there is a lot of snowfall and it's a difficult problem to overcome. Most fungicides won't work on it.

The solution is to aerate the soil and improve the drainage. Try not to walk over the area because you can spread the fungus from one place to another. Apply 1 cm (3/8 inch) of sand and feed the area with 21-0-0 fertilizer. It often helps to change the conditions of the soil by applying lime to bring up the pH.

Growing grass under evergreens

Evergreen trees are incompatible with lawns for a number of reasons. The branches cut down on light reaching the grass, the tree roots suck water out of the ground and the needles boost the acidity of the soil to make it unfriendly to grass.

One solution is to prune the offending tree back a wee bit. The height of the tree is not so much a problem as its width. Start by taking branches out at the bottom of the tree and work upwards to give the tree a more narrow form. The tree can look far more attractive and shade the lawn less.

During the summer a large evergreen tree will pull about 380 litres (100 gallons) of water out of the soil every day. The grass is trying to compete with that. Try overseeding the area with a more shade-tolerant grass, and make sure you apply extra water to that part of the lawn. To counteract the acidity created by the falling needles, apply dolomite lime underneath the tree every fall.

Other Groundcovers

There are many situations in our landscapes where lawns may not be the most appropriate groundcover. Steep slopes, small areas that are awkward to mow, very shady spots, underneath big trees, and extremely wet soils are some "non-lawn" locations.

There are all kinds of alternatives for a green covering, but consider a few things before you start planting.

- Maintenance should be minimal. You don't want to be continually pruning, watering and fertilizing an area that is supposed to be low maintenance.
- The groundcover you select should be compatible with the area. All the awkward lawn spots I just mentioned have an appropriate plant solution.
- Take time to plant your groundcover properly, with good soil, nutrients and regular watering until it's established. Weed control is really important at this stage. It's a good idea to prune groundcovers back two or three times as young plants so they will get really good and full before they take off. Start with well-established 10-cm (4-inch) pots for good economy and space them in a staggered fashion about 45 to 60 cm (18 to 24 inches) apart.

A few fast lawn alternatives

Shady areas

Baltic ivy (*Hedera helix* 'Baltica'). Zone 5. Grows to 15 cm (6 inches) high, with a 3-m (10-foot) spread. This is one of the hardiest ivies. It has small, dark green, glossy leaves with white veins.

Japanese spurge (*Pachysandra terminalis*). Zone 3. Grows to 20 cm (8 inches) high; 90-cm (3-foot) spread. This is one of the best evergreen groundcovers for shade. It has white flower spikes in spring and beautiful glossy green leaves for a very rich-looking appearance.

Salal (*Gaultheria shallon*). Zone 5. Grows to 50 cm (20 inches) high, with a 90-cm (3-foot) spread. This native of the Pacific coast has deep green foliage and small white to pink flowers, followed by blue-black fruits. Great to plant under trees.

Steep slopes
Cotoneaster dammeri. Zone 6. Grows 10 to 15 cm (4 to 6 inches) high, with a 2-m (7-foot) spread. One of the most elegant groundcovers, with a low, flat appearance and attractive overall look. White flowers in spring are followed by red, fall through winter berries.
- *C. dammeri* 'Lowfast' is a zone 6 plant that grows 30 to 50 cm (12 to 20 inches) high, with a 3-m (10-foot) spread. It's very fast-growing and will root where it touches the soil to help hold moderate slopes. Small leaves, with white flowers in spring and red berries in fall and winter.

Kinnikinnick (*Arctostaphylos uva-ursi* 'Vancouver Jade'). Zone 2. Up to 4 cm (1½ inches) high, with a 3-m (10-foot) spread. Small, shiny, dark green leaves turn bronze in winter. Beautiful clusters of pink heatherlike flowers in spring. Likes well-drained, sandy soil. Spreads uniformly and quickly.

Wet areas
Cranberry (*Vaccinium macrocarpon*). Zone 3. Grows to 10 cm (4 inches) high, 90-cm (3-foot) spread. Small, dark green leaves turn brilliant bronze in fall and winter. Tiny, white flowers produce huge edible berries in fall. Cranberry is one of the most underutilized groundcovers.

Dwarf Chinese astilbe (*Astilbe chinensis* 'Pumila'). Zone 3. Grows to 15 cm (6 inches) high, with a 60-cm (24-inch) spread. Compact, dull green, thick foliage and lavender-pink flowers July through October. We have it edging a stream at Minter Gardens, and it looks fabulous.

Palm sedge (*Carex muskingumensis*). Zone 4. A grass suitable for wet areas. (See chapter 6, "Perennial ornamental grasses and grasslike plants," for more details.)

Under trees
Japanese spurge. (See "Shady areas" above.)

Laurel (*Prunus laurocerasus* 'Otto Luyken'). Zone 6. Grows to 60 cm (24 inches) high, with a 2-m (7-foot) spread. Dark green, narrow leaves. Very compact, with white flowers in spring, producing black fruits in fall.

Salal. (See "Shady areas" above.)

Between stepping stones
Woolly thyme (*Thymus pseudolanuginosus*). Zone 2. Grows to 6 cm (2.5 inches) high, with a 50-cm (20-inch) spread. One of the most uplifting thymes, with silver-grey foliage that forms a dense mat with pinkish flowers.

There are hundreds of groundcovers that will work as an alternative to lawns. The ones listed here meet all the criteria for a fast, easy and attractive groundcover. (See chapter 6, "Perennial ornamental grasses and grasslike plants—Groundcovers," for more great ideas.)

Wildflowers
Many folks imagine the beauty of a wildflower meadow when they consider wildflowers as an alternative to lawns. Wildflowers can in no way replace a lawn, but by blending in a grass such as hard fescue, you will have a green base in winter, with colour during the flowering season. Keep in mind wildflowers do not always look good and you will have to overseed from time to time. It took Mother Nature centuries to get it right, so have patience. Try it on a small scale for a couple of years, then make a decision about expanding. Be aware that there are various wildflower mixes available for all types of climates, and it is important to get some advice and top-quality seed for your particular area.

Sow the wildflowers along with some hard fescue grass seed. The hard fescue stays low and compact. It's used in very low-maintenance mixes and it will give you some green even when there are no flowers in bloom. You want some flowers to bloom early, others in mid-season and some as late as possible. Remember to allow the seeds of annuals and biennials to mature, drop and recycle on their own.

If your topsoil is a bit heavy, you will want to create a sandy, lighter soil. Work that soil by adding some bark mulch. If it's full of gravel, put some compost into the mix. As soon as all danger of frost is past in the spring, get the seeds in. Timing is crucial because there is going to be a race between the natural weeds and your seeds, so you want the seeds to get the best opportunity.

Remember, don't fertilize wildflower areas to promote growth. You want low nutrient levels for shorter plants with more flowers.

It is hit and miss when it comes to germination, and you will have to experiment with different mixes to see which works best for your particular spot. There are coastal, alpine, arid and northern mixes.

Moss

Most mosses will grow naturally in a moist, shady location, and in the right spot look very attractive. True mosses spread by spores shed by existing plants. Also good for the moss look is a low-growing plant called Irish moss (*Sagina subulata*) and its chartreuse form, Scotch moss (*S. subulata* 'Aurea'). True mosses are not available in garden centres, although these plants are.

They need a well-drained sandy loam. When you plant them, cut them in 2.5-cm (1-inch) squares. Place them about 15 cm (6 inches) apart and you will have them growing together in a solid mat in one season. Be sure to roll this area much as you would a lawn to keep the plants from pushing up in the centre like a mushroom. You may have to do this every couple of years.

Encouraging moss to grow on stones

True mosses can be difficult to cultivate, so if you're covering a lot of ground, try the following tip to make a moss-friendly environment. Boil some rice in water, strain out the rice and pour the water where you want the moss to grow. This creates an ideal environment for moss spores to become established.

Index

Abelia (*Abelia* hybrids) 139, 142
Abeliophyllum distichum (white forsythia) 137
Abies koreana (Korean fir) 158
Abutilon species 70; *A. megapotamicum* (flowering maple) 70; *A. pictum* 70
Acer species (maples) 131–32; *A. griseum* (paperbark maple) 132; *A. palmatum* (Japanese maple) 131; *A. rubrum* (red maple) 131
Achillea (yarrow) 94
Aconite, winter (*Eranthus hyemalis*) 197, 199
Acorus species (sweet flag): *A. calamus* 98; *A. gramineus* 82
Actinidia varieties (kiwi) 189–91; *A. kolomikta* (ornamental kiwi) 119
Adam's needle (*Yucca filamentosa*) 149–50
Aerating lawns 203–4
Ageratum 63
Ajuga species (bugleweed) 82, 96; *A. reptans* 96
Akebia, five-leafed (*Akebia quinata*) 118
Alchemilla mollis (lady's mantle) 95
Allium species (chives, leeks): *A. moly* (lily leek) 198; *A. schoenoprasum* (chives) 53, 55; *A. tuberosum* (garlic chives) 53
Almond (*Prunus dulcis*) 178
Aloysia triphylla (lemon verbena) 52
Alyssum (*Labularia maritima*) 63, 80
Amaryllis belladona 58
Amelanchier alnifolia (Saskatoon berry) 187
Ampelodesmos mauritanica (Mauritania vine reed) 99
Ampelopsis, porcelain (*Ampelopsis brevipendunculata*) 119
Anemone (*Anemone* varieties) 58; Japanese 91; windflower 198

Anethum graveolens (dill) 52
Anise (*Pimpinella anisum*) 51, 56
Anise hyssop 57
Annuals 59–70
Antirrhinum majus (snapdragon) 64, 69
Aphids 46
Apple (*Malus domestica* varieties) 170–73; apple scab 179; columnar 172–73; rootstock 164; storing 170
Apricot (*Prunus armeniaca*) 173, 179; rootstock 164
Arabis (rockcress) 86–87
Arbutus unedo (strawberry tree) 151
Arctostaphylos uva-ursi (kinnikinnick) 214
Argeranthemus frutescens (marguerite daisy) 64, 69, 74
Aristolochia macrophylla (Dutchman's pipe) 118
Arrhenatherum elatius (bulbous oat grass) 100–101
Artemisia (*Artemisia* species): French tarragon 53; wormwood 75, 95
Aruncus dioicus (goatsbeard) 88
Arundo donax (giant reed grass) 99
Asparagus 31–32; asparagus beetle 32
Asparagus fern (*Asparagus densiflorus*) 76
Aster, China (*Callistephus chinensis*) 62, 64
Astilbe, dwarf Chinese (*Astilbe chinensis*) 96–97, 214
Athyrium nipponicum (Japanese painted fern) 91
Autumn crocus (*Colchicum autumnale*) 200
Azaleas 13, 17, 58, 137–38, 140, 152–53

Baby's breath (*Gypsophila paniculata*) 94
Bachelor's buttons 58

Bacillus thurengiensis (B.T.) 46
Bacopa (*Sutera cordata*) 74
Barrenwort (*Epimedium* × *rubrum*) 96
Basil (*Ocumum basilicum*) 47, 50, 51
Bay laurel (*Laurus nobilis*) 51
Beans 13, 28, 30, 32–33, 58
Beautyberry (*Callicarpa bodinieri*) 142
Beautybush (*Kolkwitzia amabilis*) 138
Bedding plants 59–70
Bee balm (*Monarda didyma*) 54–55
Beech (*Fagus sylvatica*) 131
Beets 29, 33
Begonias (*Begonia* varieties): fibrous 63, 64, 66, 75–76; tuberous 66
Bellis perennis (English daisy) 69
Bergenia (*Bergenia* hybrids) 91, 94–95
Berries 181–85
Betony (*Stachys officinalis*) 55
Betula utilis var. *jacquemontii* (Himalayan birch) 135
Bidens aurea 74–75
Birch, Himalayan (*Betula utilis* var. *jacquemontii*) 135
Blackberries (*Rubus fruticosus*) 182
Black-eyed susan (*Rudbeckia fulgida*) 92
Black spot 86
Bleeding heart (*Dicentra formosa*) 87
Blight 26, 42, 47
Bluebell, English (*Hyacinthoides non-scripta, Scilla non-scripta*) 198
Blueberry (*Vaccinium* varieties) 180, 184–85; creeping 191–92
Blue-green rush (*Juncus inflexus*) 100
Blue oat grass (*Helictotrichon sempervirens*) 100
Bone meal 14
Borage (*Borago officinalis*) 56, 58

Borders 61
Boron 14, 179
Boston ivy (*Parthenocissus tricuspidata*) 118
Bouteloua species: *B. curtipendula* (sideoats grass) 99; *B. gracilis* (mosquito grass) 97–98, 99
Boysenberries 184
Brachycome (Swan River daisy) 76, 77
Brassica oleracea (ornamental cabbage, kale) 70
Brassicas 30, 33–34
Broadleaf evergreens 147–58, 191–92; for berries 148; calendar of 147–52
Broccoli 29, 33–34
Broom (*Cytisus* species) 139–40
Broom, creeping (*Genista pilosa*) 149
Browallia (*Browallia speciosa*) 66
Brown-eyed susan (*Rudbeckia* hybrids) 64, 65
Brussels sprouts 34
Buartnut 178
Buddleia species (butterfly bush): *B. davidii* 139, 140; *B. nanhoensis* (dwarf butterfly bush) 141
Bugleweed (*Ajuga reptans*) 96
Bulbous oat grass (*Arrhenatherum elatius*) 100
Bulbs 193–202; care 193–95, 196–97; disease and pest control, 202; forcing 201–2; naturalized 197; planting 193–94, 196
Bush fruits 184–87
Butterfly bush (*Buddleia* species) 139, 140, 141

Cabbage (*Brassica* species) 29, 34; ornamental 69, 70
Calceolaria (slipperflower) 75
Calcium 14
Calendula 56, 62; *C. officinalis* (pot marigold) 56
Callicarpa bodinieri (beautyberry) 142
Callistephus chinensis (China aster) 64
Calluna vulgaris (Scots heather) 150
Camellias 13

Campsis × *tagliabuana* (trumpet creeper) 120
Candytuft (*Iberis sempervirens*) 86–87, 152
Cane fruits 182–84
Canker 179
Carex species (sedge): *C. buchananii* (leatherleaf sedge) 82, 98, 101; *C. comans* (New Zealand hair sedge) 101; *C. elata* 91; *C. hachijoensis* 82; *C. morrowii* (Japanese sedge) 98, 100; *C. muskingumensis* (palm sedge) 100, 214; *C. pseudocyperus* (cyperus sedge) 100
Carnations (*Dianthus* species) 58, 63–64, 81, 96
Carrot rust fly 47
Carrots 28, 29, 30, 34
Castanea mollisima (chestnut) 178
Castor bean plant (*Ricinus communis*) 69
Catmint, variegated (*Plectranthus fors*) 76
Catnip (*Nepeta cataria*) 54
Cattail, miniature (*Typha minima*) 101
Cauliflower 29, 34
Ceanothus thyrsiflorus (California lilac) 149
Cedar (*Thuja* varieties) 160, 161
Cedar, weeping blue Atlas (*Cedrus atlantica* 'Glauca Pendula') 159
Celeriac 30
Celery 30
Centradenia inaequilateralis 76
Cercidiphyllum japonicum (katsura) 130–31
Cercis canadensis (eastern redbud) 132
Chaenomeles japonica (flowering quince) 137, 141
Chamaecyparis species (cypress): *C. nootkatensis* 'Pendula' (weeping nootka cypress) 159; *C. obtusa* (hinoki cypress) 159, 160; *C. pisifera* (gold thread cypress) 160
Chamomile, Roman (*Chamaemelum nobile*) 54
Chenopodium album (lamb's quarters) 46
Cherry (*Prunus* varieties) 133–34, 173–74, 179; birchbark 135;

Japanese flowering 133–34; pruning 167; rootstock 164; sour 174; sweet 173–74
Cherry, Cornelian (*Cornus mas*) 137
Chestnuts (*Castanea* varieties) 178
Chimonanthus praecox (wintersweet) 137
Chinese lantern plant (*Physalis alkekengii*) 94
Chionodoxa luciliae (glory-of-the-snow) 198, 202
Chives (*Allium* species) 50, 53, 55, 58
Chlorine 14
Chrysanthemum frutescens (marguerite daisy) 64, 69, 74
Cilantro (*Coriandrum sativum*) 51–52
Cinquefoil, shrubby (*Potentilla fruticosa*) 138–39, 142
Claret vine (*Vitis vinifera*) 118–19
Clematis (*Clematis* species) 120–26; *C. alpina* 121; *C. armandii* 121; *C. chrysocoma* 122; *C. florida* 121; *C. heracleifolia* 120; *C. integrifolia* 120; *C. macropetala* 121; *C. montana* 121–22; *C. tangutica* 121, 122; *C. terniflora* (sweet autumn clematis) 122; *C. vitalba* 122; clematis wilt 125; pruning 124–26
Cleome hassleriana (spider flower) 65
Climbers 117 26; care 117–18
Cockscomb 64
Colchicum autumnale (autumn crocus) 200
Cold frame, 22
Coleus (*Solenostemon*) 66, 70, 75
Colour planning 59–61
Companion planting 46–47, 57
Compost 14–17
Conifers 17, 158–61; pruning 159
Consolida ambigua (larkspur) 64
Container gardening 27–29, 71–83; fall and winter 81–83; feeding 73; herbs 50, 55; planting 72–73; vegetables 27–29, 41, 44; watering 73
Convallaria majalis (lily-of-the-valley) 96

Copper 14
Coral bells (*Heuchera*) 95
Cordyline australis (dracaena palm) 70
Coreopsis (*Coreopsis* species): pink-flowered coreopsis 92–93; tickseed 93–94
Coriander, leaf (*Coriandrum sativum*) 51–52
Coriander, Vietnamese (*Polygonum odoratum*) 52
Corm 193
Corn 35
Cornelian cherry (*Cornus mas*) 137
Cornus species (dogwood) 81, 132–33, 142; *C. atrosanguinea* 'Sibirica' (red twig dogwood) 142; *C. florida* (eastern dogwood) 133; *C. kousa* (Japanese dogwood) 132–33; *C. kousa* var. *chinensis* (Chinese dogwood) 132–33; *C. mas* (Cornelian cherry) 137; *C. sanguinea* 81, 142; *C. stolonifera* 'Flaviramea' (yellowtwig dogwood) 142
Corydalis, yellow (*Corydalis lutea*) 88
Corylus avellana 'Contorta' (Harry Lauder's walking stick) 135, 139
Corylus maxima (filbert) 178
Cosmos (*Cosmos bipinnatus*) 62, 64, 65
Cotoneaster (*Cotoneaster* species) 151, 214
Crabapple (*Malus* varieties) 132, 174
Cranberry (*Vaccinium* species) 191, 214; mountain 192
Cranberry, highbush (*Viburnum trilobum*) 187
Cranesbill, bloody (*Geranium sanguineum*) 92
Creeping jenny (*Lysimachia nummularia*) 74, 76, 82, 96
Crocus (*Crocus* species) 197, 198, 200, 201; banded 200; Dutch 198, 201; saffron 55–56, 200; showy autumn 200
Crocus, autumn (*Colchicum autumnale*) 58, 200

Cucumbers 27, 30, 35–36
Cupressocyparis × *leylandii* (Leyland cypress) 161
Currants (*Ribes* species) 185–86; flowering 137
Cuttings, growing from 68
Cyclamen (*Cyclamen* species) 199
Cypress (*Chamaecyparis* species): gold thread 160; hinoki 159, 160; weeping nootka 159
Cypress, Leyland (*Cupressocyparis* × *leylandii*) 161
Cytisus species (broom): *C. kewensis* 140; *C. praecox* 139; *C. purgans* 140; *C. scoparius* 139

Daffodil, winter (*Sternbergia lutea*) 201
Daffodil (*Narcissus*) 197, 198, 200, 201
Daisies: African (*Dimorphotheca*) 62, 64, 65; English (*Bellis perennis*) 69; Livingstone (*Mesembryanthemum*) 62; marguerite (*Chrysanthemum frutescens, Argeranthemus frutescens*) 64; Swan River (*Brachycome* species) 76, 77
Daphne (*Daphne* species) 137, 149; February 137; rock 149
Daylily 58, 92
Dead nettle (*Lamium maculatum*) 76
Delphiniums 86
Deschampsia caespitosa (tufted hair grass) 100
Deutzia (*Deutzia gracilis*): dwarf 141; slender 139
Dianthus species (carnations, pinks) 58, 69, 96; *D. caryophyllus* 63–64, 81
Diatomaceous earth 46
Dicentra species (bleeding heart): *D. formosa* 87; *D. spectabilis* 87
Dill (*Anethum graveolens*) 52
Dimorphotheca hybrids (African daisy, star-of-the-veldt) 62, 65
Dogwood (*Cornus* varieties) 81, 132–33, 142; bloodtwig 142; Chinese 132–33; Japanese 132–33; red twig 142; yellowtwig 142
Dracaena palms (*Cordyline australis*) 70

Dusty miller (*Senecio cineraria*) 63, 69, 70
Dutchman's pipe (*Aristolochia macrophylla*) 118

Echinacea species (purple coneflower): *E. angustifolia* 55; *E. purpurea* 93
Echinops ritro (globe thistle) 94
Edging 61
Edible flowers 49, 57–58
Eggplant 30
Elderberries (*Sambucus canadensis*) 187
Enkianthus, redvein (*Enkianthus campanulatus*) 137
Epimedium × *rubrum* (barrenwort) 96
Epsom salts (magnesium sulphate) 14
Eranthus hyemalis (winter aconite) 199
Erianthus ravennae (ravenna grass) 99
Erica species (heather): *E. carnea* (winter-flowering heather) 147; *E.* × *darleyensis* 147
Eryngium alpinum (sea holly) 94
Erysimum (wallflower) 96
Erythronium dens-canis (dogtooth violet) 199
Escholtzia californica (California poppy) 62
Espalier 168
Euonymus fortunei 118, 150
Euphorbia species (spurge) 56, 70, 82, 95; *E. amygdaloides* (wood spurge) 82, 95; *E. lathyris* (mole plant) 56; *E. marginata* (snow on the mountains) 70; *E. myrsinites* 95
Exochorda × *macrantha* (pearlbush) 137

Fagus sylvatica (beech) 131
Fairy ring 213
Fanflower (*Scaevola aemula*) 75, 77
Feather grass (*Stipa capillata*) 99
Fennel (*Foeniculum vulgare*) 53, 58
Fertilizers 14
Fescue, blue (*Festuca cinerea*) 99
Feverfew, dwarf 63

Figs (*Ficus carica*) 174
Filbert (*Corylus maxima*) 178
Fir, Korean (*Abies koreana*) 158
Firethorn (*Pyracantha* hybrids) 151
Flame grass (*Miscanthus sinensis*) 101
Flowering maple (*Abutilon* species) 70
Foamflower (*Tiarella* species) 87–88, 91
Foeniculum vulgare (fennel) 53
Forget-me-not, creeping (*Omphalodes verna*) 87
Forsythia (*Forsythia* species) 137, 141
Forsythsia, white (*Abeliophyllum distichum*) 137
Fountain grass (*Pennisetum* species) 94, 99, 101; oriental 101
Fragaria (strawberry) 97, 181–82
French tarragon (*Artemisia dracunculus* var. *sativa*) 53
Fringe cups (*Tiarella cordifolia*) 87–88
Fringe flower (*Loropetalum chinense*) 138, 141
Fritillary (*Fritillaria* species): Persian fritillary 199
Fruit trees 163–77; care 166; espalier 168; fruit drop 179; grafts 163–64; planting 165–66; pollination 164–65; pruning 166–68, 169
Fuchsias (*Fuchsia* varieties) 66, 79–80, 88; hardy 88
Fungal diseases 42, 51, 86, 179
Fusarium wilt 51

Galanthus nivalis (snowdrop) 199, 201
Garden balsam (*Impatiens balsamina*) 65, 66
Garlic 46, 56
Garrya elliptica (silk tassel) 148
Gaultheria species: *G. mucronata* (Chilean pernettya) 150–51; *G. procumbens* (wintergreen) 149; *G. shallon* (salal) 213
Gazania 64
Genista pilosa (creeping broom) 149

Geraniums (*Pelargonium* species) 56, 58, 64, 67, 77–78, 92; scented 56; trailing, or ivy 77–78; zonal 78
Geranium sanguineum (bloody cranesbill) 92
Giant reed grass (*Arundo donax*) 99
Ginkgo biloba (maidenhair tree) 131
Glechoma hederaca (nepeta) 74
Globe thistle (*Echinops ritro*) 94
Glory-of-the-snow (*Chionodoxa luciliae*) 198, 201
Goatsbeard (*Aruncus dioicus*) 88
Golden ray (*Ligularia dentata*) 88
Goldenrod (*Solidago odora*) 55
Goniolimon tataricum (Tatarian statice) 94
Gooseberries (*Ribes uva-crispa* var. *reclinatum*) 187
Gourds 39
Grape hyacinth (*Muscari* species) 199, 200, 201
Grapes (*Vitis vinifera*) 188–89; ornamental 118–19
Grape vines 180
Grasses, ornamental 97–102
Grass mixes 209–10
Groundcovers 95–97, 203–15
Gypsophila paniculata (baby's breath) 94

Hakone grass (*Hakonechloa macra*) 100–101
Hamamelis mollis (Chinese witch-hazel) 137
Hanging baskets 71–83
Harry Lauder's walking stick (*Corylus avellana* 'Contorta') 135, 139
Heartnut 178
Heather (*Erica* species) 13, 17, 147, 150; winter-flowering 147
Heather, Scots (*Calluna vulgaris*) 150
Hedera species (ivy): *H. algeriensis* (Algerian ivy) 150; *H. colchica* (Persian ivy) 118; *H. helix* (English ivy) 74, 119; *H. helix* 'Baltica' (Baltic ivy) 213
Hedges 160–61
Helenium autumnale (common sneezeweed) 92

Helichrysum species: *H. bracteatum* (strawflower) 62, 77; *H. petiolare* (licorice vine) 75; *H. thianschanicum* 82
Helictotrichon sempervirens (blue oat grass) 100
Heliotrope (*Heliotropium arborescens*) 66, 81
Hellebore, stinking (*Helleborus foetidus*) 95
Hemerocallis 'Stella d'Oro' (daylily) 92
Hemlock (*Tsuga* varieties): low spreading 160; sargent's weeping 158
Herbicides 211–12
Herbs 49–58; in containers 50, 55; culinary 51–54; drying 50–51; freezing 50; growing indoors 55; ornamental 54–56
Heuchera (coral bells) 95
Hibiscus syriacus (rose-of-sharon) 139, 140
Highbush cranberries (*Viburnum trilobum*) 187
Holly (*Ilex* varieties) 151–52
Honeysuckle (*Lonicera* species) 119
Honeysuckle, Himalayan (*Leycesteria formosa*) 150
Hop vine (*Humulus lupulus*) 191; ornamental 118
Hostas (*Hosta* varieties) 88–91
Humulus lupulus (hop vine) 118, 191
Hyacinth, common (*Hyacinthus orientalis*) 199, 201
Hyacinth, grape (*Muscari* species) 199, 200, 201
Hyacinthoides non-scripta (English bluebell) 198
Hydrangeas (*Hydrangea* species) 58, 61, 139–41; climbing 118; hills-of-snow 140
Hypericum patulum (St. John's wort) 139, 141

Iberis sempervirens (candytuft) 86–87, 152
Ilex (holly); *I. aquifolium* (English holly) 151; *I.* × *meserveae* (blue holly) 151–52; *I. verticillata* (winterberry) 142

Impatiens (*Impatiens* varieties) 58, 59, 63, 65, 66, 78; garden balsam 65, 66; New Guinea impatiens 78
Imperata cylindrica (Japanese blood grass) 101
Insect control 45–47, 107
Iris (*Iris* species) 55, 58, 88, 200, 201
Iron 14
Itea virginica (sweetspire) 139, 142
Ivy (*Hedera* species): Algerian 150; Baltic 215; English 74, 119; Persian 118, 119
Ivy (*Parthenocissus* species) 118, 119

Jacob's ladder (*Polemonium caeruleum*) 88
Japanese blood grass (*Imperata cylindrica*) 101
Japanese silver grass (*Miscanthus sinensis*) 99, 101
Japanese spurge (*Pachysandra terminalis*) 97
Jasmine (*Jasminum* species) 120
Juglans species (walnut): *J. nigra* (black walnut) 178; *J. regia* (Carpathian walnut) 179
Juncus species (rush) 100
Juniper (*Juniperus* species): blue star 159–60; creeping 160; hollywood 158

Kale, ornamental (*Brassica oleracea*) 69, 70
Kalmia latifolia (mountain laurel) 149, 150
Katsura (*Cercidiphyllum japonicum*) 130–31
Kerria japonica 137
Kinnikinnick (*Arctostaphylos uva-ursi*) 214
Kiwi (*Actinidia* varieties) 189–91; ornamental 119
Knotweed, Himalayan (*Polygonum affine*) 97
Kolkwitzia amabilis (beautybush) 138

Labularia maritima (alyssum) 63, 80
Ladybugs 46

Lady's mantle (*Alchemilla mollis*) 95
Lamb's ears (*Stachys byzantina*) 95
Lamb's quarters (*Chenopodium album*) 46
Lamium species (nettle): *L. galeobdolon* (silver nettle vine, yellow archangel) 74, 76, 97; *L. maculatum* (variegated dead nettle) 76, 82
Larkspur (*Consolida ambigua*) 58, 64
Lathyrus odoratus (sweet peas) 65
Laurel (*Prunus laurocerasus*) 214
Laurel, mountain (*Kalmia latifolia*) 149, 150
Laurus nobilis (bay laurel) 51
Lavender, English (*Lavandula angustifolia*) 54, 58, 96
Lavender cotton (*Santolina chamaecyparissus*) 54
Lawns 203–13; aerating 203–4; fertilizing 204–5, 206; mowing 207–8, 209; new 208–10; overseeding 207; problems with 208, 211–13; watering 206–7
Leek, lily (*Allium moly*) 198
Leeks 29, 30
Lemon balm (*Melissa officinalis*) 53
Lemon verbena (*Aloysia triphylla*) 52
Lettuce 28, 29, 30, 36–37
Leucojum vernum (spring snowflake) 200
Leucothöe fontanesiana 150
Levisticum officinale (lovage) 53
Lewisia (*Lewisia cotyledon*) 92
Leycesteria formosa (Himalayan honeysuckle) 150
Licorice vine (*Helichrysum petiolare*) 75, 76
Ligularia dentata (golden ray) 88
Lilac: California 149; dwarf Korean 138, 141; French 138
Lilies (*Lilium* species) 199, 200
Lily-of-the-valley (*Convallaria majalis*) 58, 96
Lily-of-the-valley shrub (*Pieris japonica*) 148, 150
Lime 12–13, 205–6
Limonium species (statice) 64, 94

Linaria maroccana (toadflax) 69
Linarias 59
Lingonberry (*Vaccinium vitis-idaea* var. *minus*) 192
Liquidambar styraciflua (sweet gum) 131
Liriodendron tulipifera (tulip tree) 131
Lobelia (*Lobelia* species) 58, 63, 66, 75
Loganberries 184
London plane tree (*Platanus acerifolia*) 131
Lonicera species (honeysuckle) 119
Loosestrife, yellow (*Lysimachia punctata*) 88
Loropetalum chinense (fringe flower) 138, 141
Lotus vine 75
Lovage (*Levisticum officinale*) 53
Lungwort (*Pulmonaria* species) 88, 91
Lupine 58, 86
Luzula nivea (snowy woodrush) 100
Lysimachia species: *L. congestiflora* 75; *L. nummularia* (creeping jenny) 74, 76, 82, 96; *L. punctata* (yellow loosestrife) 88

Maggots 46
Magnesium 13, 14
Magnesium sulphate 14
Magnolia (*Magnolia* species) 134
Mahonia species: *M. aquifolium* (Oregon grape) 149, 150; *M.* × *media* 149, 150; *M. nervosa* 149; *M. repens* 149, 150
Maiden grass (*Miscanthus sinensis*) 94
Maidenhair tree (*Ginkgo biloba*) 131
Malus domestica (apples) 170–73
Manganese 14
Manure 11, 17
Maple (*Acer* species): Japanese 131, 132; paperbark 132; red 131
Marguerite (*Chrysanthemum frutescens*, *Argeranthemus frutescens*) 69, 74

Marigold: dwarf 63; pot, or English 56, 62; Scotch 58; triploid 64
Marjoram, sweet (*Origanum majorana*) 50, 52
Matthiola species (stocks) 65, 69
Mauritania vine reed (*Ampelodesmos mauritanica*) 99
Maypop (*Passiflora incarnata*) 120
Meconopsis betonicifolia 85
Melissa officinalis (lemon balm) 53
Melons 30
Mentha species (mints) 53–54
Mesembryanthemum (Livingstone daisy) 62
Micronutrients 14
Mildew 26, 86
Mimulus (monkey flower) 66, 69, 74
Mints (*Mentha* species) 50, 53–54
Miscanthus sinensis (Japanese silver grass) 94, 99, 101, 102
Mock orange (*Philadelphus virginalis*) 138, 141
Mole plant (*Euphorbia lathyris*) 56
Molybdenum 14
Monarda didyma (bee balm) 54–55
Mondo grass (*Ophiopogon planiscapus*) 101
Monkey flower (*Mimulus* hybrids) 66, 69, 74
Monkshood 58
Morus bombycis (contorted mulberry) 135
Mosquito grass (*Bouteloua gracilis*) 97–98, 99
Mosquito plant (*Pelargonium citrosa* 'Van Leenii') 56
Moss (*Sagina subulata*) 214
Moss control 212
Moths 46
Mountain laurel (*Kalmia latifolia*) 149, 150
Mowing lawns 207–8, 209
Mud slurry 22
Mulberry, contorted (*Morus bombycis* 'Unryu') 135
Mulch 18, 86
Muscari species (grape hyacinth) 199, 200, 201
Mushroom control 212
Myrtle (*Myrtus communis*) 56

Narcissus 198, 200, 201
Nasturtium (*Tropaeolum majus*) 46, 56, 58, 62
Nectarine (*Prunus persicus* var. *nectarine*) 174–75; rootstock 164
Nematodes 46, 89
Nemesia denticulata 74, 75, 77, 81
Nepeta (*Glechoma hederacea*) 74
Nepeta cataria (catnip) 54
Nicotiana alata 66
Nitrogen 11, 13, 14, 15, 27
Nut trees 177–79

Oak, pin (*Quercus palustris*) 131
Ocumum basilicum (basil) 51
Omphalodes verna (creeping forget-me-not) 87
Onions 29, 30, 40
Ophiopogon planiscapus (mondo grass) 101
Oregon grape (*Mahonia* hybrids) 149, 150
Oregano (*Origanum vulgare*) 54
Organic insect and pest control 45–47, 107
Origanum species: *O. majorana* (sweet marjoram) 52; *O. vulgare* (oregano) 54
Orris (*Iris germanica* var. *florentina*) 55
Oxydendrum arboreum (sourwood) 135

Pachysandra terminalis (Japanese spurge) 97, 213
Painted fern, Japanese (*Athyrium nipponicum*) 91
Painted tongue (*Salpiglossis sinuata*) 65
Panicum virgatum (switch grass) 101, 102
Pansies (*Viola* × *wittrockiana*) 58, 69, 74
Parsley (*Petroselinum crispum*) 47, 50, 54
Parsley, Chinese (*Coriandrum sativum*) 51–52
Parthenocissus species (ivy): *P. quinquefolia* (Virginia creeper) 119; *P. tricuspidata* (Boston ivy) 118
Passion flower, wild (*Passiflora incarnata*) 120

Peach (*Prunus persicus*) 175–76, 179; rootstock 164
Pear (*Pyrus communis*) 176–77; rootstock 164
Pear, oriental (*Prunus pyrifolia*) 175
Pearlbush (*Exochorda* × *macrantha*) 137
Peas 28, 30, 37
Peat moss 13
Pelargonium hybrids (geranium) 56, 77–78
Pennisetum species (fountain grass) 94, 98, 99, 100, 101
Peonies 86, 87
Peppers 28, 30, 37–38
Perennials 85–102; dividing 86
Periwinkle (*Vinca* species) 74, 76
Pernettya, Chilean (*Pernettya mucronata*, *Gaultheria mucronata*) 150–51
Persian ivy (*Hedera colchica*) 118
Pest control 45–47, 107
Petroselinum crispum (parsley) 54
Petunia (*Petunia* varieties) 58, 64; trailing 76–77, 81
pH 12–13
Phalaris arundinacea (ribbon grass) 100
Philadelphus hybrids (mock orange) 138, 141
Phlox (*Phlox* species) 93; summer 93, 96
Phosphorus 14
Physalis alkekengii (Chinese lantern plant) 94
Picea species (spruce): *P. glauca albertiana* (Alberta spruce) 159; *P. glauca* 'Pendula' (weeping white spruce) 159; *P. omorika* 'Nana' (dwarf Serbian spruce) 160; *P. pungens* 'Glauca Globosa' (dwarf globe blue spruce) 160
Pieris japonica (lily-of-the-valley bush) 148, 150
Pimpinella anisum (anise) 51, 56
Pincushion flower (*Scabiosa columbaria*) 92
Pine, dwarf mugo (*Pinus mugo* var. *pumilo*) 160
Pinks (*Dianthus* species) 69, 96
Plane tree, London (*Platanus acerifolia*) 131

Planting seeds 20–21
Planting times 30
Platanus acerifolia (London plane tree) 131
Plectranthus fors (variegated catmint) 76
Plum (*Prunus* varieties) 164, 177
Plum, flowering (*Prunus cerasifera*) 135
Poisonous plants 58
Polemonium caeruleum (Jacob's ladder) 88
Polygonum species: *P. affine* (Himalayan knotweed) 97; *P. aubertii* (silver lace vine) 119; *P. odoratum* (Vietnamese coriander) 52
Poppy, California (*Escholtzia californica*) 62
Porcelain ampelopsis (*Ampelopsis brevipendunculata*) 119
Porcupine grass (*Miscanthus sinensis* 'Strictus') 102
Portulaca 62, 64
Potash 14
Potassium 14
Potatoes 13, 38–39, 47
Potentilla fruticosa (shrubby cinquefoil) 138–39, 142
Prairie cord grass (*Spartina pectinata*) 100
Primula vialii 85
Prunus species: *P. armeniaca* (apricots) 173; *P. avium* (sweet cherries) 173–74; *P. cerasifera* (flowering plum) 135; *P. cerasus* (sour cherries) 174; *P. domestica* (European plum) 177; *P. dulcis* (almond) 178; *P. laurocerasus* (laurel) 150, 216; *P. persicus* (peaches) 175–76; *P. persicus* var. *nectarine* (nectarines) 174–75; *P. pyrifolia* (oriental pears) 175; *P. saliciana* (Japanese plum) 177; *P. serrula* (birchbark cherry) 135; *P. subhirtella* (flowering cherry) 133–34
Pulmonaria species (lungwort) 88, 91
Pumpkins 30, 39–40
Purple coneflower (*Echinacea* species) 55, 93
Pyracantha (firethorn) 151
Pyrus communis (pear) 176–77

Quercus palustris (pin oak) 131
Quince, Japanese flowering (*Chaenomeles japonica*) 137, 141

Raised beds 15, 27
Raspberries (*Rubus idaeus*) 183–84
Ravenna grass (*Erianthus ravennae*) 99
Redbud, eastern (*Cercis canadensis*) 132
Rejuvenation Mix 12, 27
Rhizomes 193
Rhododendrons 13, 17, 58, 153–58; care 153–55; pruning 154; winter (*Rhododendron* × *praecox*) 148
Ribbon grass (*Phalaris arundinacea*) 100
Ribes species (currants) 185–86; flowering (*R. sanguineum*) 137; gooseberry (*R. uva-crispa* var. *reclinatum*) 187
Ricinus communis (castor bean plant) 69
Robinia, contorted (*Robinia pseudoacacia* 'Tortuosa') 135
Rockcress (*Arabis*) 86–87
Rodgersia (*Rodgersia podophylla*) 95
Rosa species (rose): *R.* × *alba* (alba rose) 105–6; *R. banksiae* (Lady Banks' rose) 104; *R.* × *centifolia* (centifolia rose) 105–6; *R.* × *centifolia muscosa* (moss rose) 106; *R. chinensis* (China rose) 106, 107, 109; *R. gallica* 104–5; *R. gigantea* 107; *R. glauca* (*R. rubrifolia*) 104; *R. multiflora* 109; *R. pimpinellifolia* (burnet rose, Scotch rose) 107; *R. rugosa* (shrub rose) 111, 142; *R. sericea* f. *pteracantha* 104
Rose-of-sharon (*Hibiscus syriacus*) 139, 140
Rosemary (*Rosmarinus officinalis*) 50, 52
Roses (*Rosa* species) 58, 103–16: alba 105; bourbon 106; burnet 107; care 114–16; centifolias 105; China 106; climbing 111–12, 113; controlling

disease and insects 107, 114, 115; damasks, 105; David Austin 110; floribunda 108–9; gallica 104–5; grandiflora 109; groundcover 112–13; hybrid tea 108; miniature 113–14; modern 108; moss 106; noisette 107; old garden 104–8; planting 114; polyantha 109–10; Portland 107; pruning 115–16; Scotch 107; shrub 110–11, 142; species 103–4; tea 107–8
Rosmarinus officinalis (rosemary) 52
Rubus species: *R. fruticosus* (blackberries) 182; *R. idaeus* (raspberries) 183–84
Rudbeckia species 64, 65; *R. fulgida* (black-eyed susan) 92
Rush: banded zebra (*Schoenoplectus tabernaemontanus*) 99; blue-green (*Juncus inflexus*) 100; spiral (*Juncus effusus*) 100
Rye 12

Saffron crocus (*Crocus sativus*) 55–56
Sage (*Salvia* species) 54, 64, 82; pineapple (*S. elegans*) 56, 58
Sagina subulata (Irish moss, Scotch moss) 214
Salal (*Gaultheria shallon*) 213
Salix species (willow): *S. matsudana* 'Tortuosa' (corkscrew willow) 135; *S. tortuosa* 82
Salpiglossis sinuata (painted tongue) 65
Salvia species (sage) 64: *S. elegans* (pineapple sage) 56; *S. officinalis* 54, 82
Sambucus canadensis (elderberry) 187
Santolina species 50; *S. chamaecyparissus* (lavender cotton) 54
Sanvitalia procumbens (creeping zinnia) 75
Sarcococca species (sweet box) 148: *S. hookeriana* var. *humilis* 148, 150; *S. ruscifolia* 148
Saskatoon berries (*Amelanchier alnifolia*) 187
Savoury (*Satureja* species): summer (*S. hortensis*) 52, 54; winter (*S. montana*) 54